BULLETIN
of the
AMERICAN MATHEMATICAL SOCIETY

VOLUME 72, NO. 1, PART II

Norbert Wiener
1894-1964

Published by the American Mathematical Society
PROVIDENCE, RHODE ISLAND
1966

Edited by

FELIX BROWDER   E. H. SPANIER

MURRAY GERSTENHABER

The BULLETIN OF THE AMERICAN MATHEMATICAL SOCIETY is published bimonthly, in January, March, May, July, September, and November. Price per annual volume is $7.00. Subscriptions, orders for back numbers, and inquiries in regard to non-delivery of current numbers should be addressed to the American Mathematical Society, P.O. Box 6248, Providence, Rhode Island 02904.

Second-class postage paid at Providence, Rhode Island and additional mailing offices. Acceptance for mailing at the special rate of postage provided for in the act of February 28, 1925, embodied in paragraph 4, Section 538, P. L. and R., authorized May 9, 1935.

Norbert Wiener

This special issue of the Bulletin of the American Mathematical Society is dedicated to the memory of Norbert Wiener in recognition of his towering stature in American and world mathematics, his remarkably many-sided genius, and the originality and depth of his pioneering contributions to science.

CONTENTS

WIENER'S LIFE

N. LEVINSON

Wiener would have been extremely gratified had he known that the American Mathematical Society would honor him in death with an issue of the Bulletin. He had had a failing heart for many years and death had come to hold no fear for him, but the possibility of lack of esteem by his colleagues was most painful to him. Unfortunately Wiener did have grave doubts all of his professional life as to whether his colleagues, especially in the United States, valued his work, and, unwarranted as these doubts were, they were very real and disturbing to him.

Wiener wrote a two-volume autobiography entitled *Ex-prodigy* and *I am a mathematician* thereby giving posterity an unusually detailed and personal picture of himself, his family, his colleagues, his times and of how he became a mathematician. It is of course an entirely honest but also entirely subjective account. Some of us come off too well, but in my opinion others are treated with a harshness that was not always warranted. Thus this article will not agree completely with Wiener's own account of the same people and events.

Norbert Wiener was born in Columbia, Missouri, November 26, 1894. His father, Leo Wiener, was a most remarkable man who had a tremendous influence on Norbert. Leo Wiener was born in Byelostok, in the ghetto area of Tsarist Russia, in 1862. He was a descendant of Aquiba Eger, Grand Rabbi of Posen from 1815 to 1837. He was also supposedly descended from Moses Maimonides. Leo Wiener's father had already broken with the ancient narrow Yiddish tradition of the ghetto and Leo was raised with literary German rather than Yiddish as his language. Whatever mixed feelings Norbert Wiener may have had as an adolescent on first realizing he was a Jew were completely gone when I met him in 1933. He was extremely proud of his scholarly ancestors and of the outstanding achievement of the Jews in mathematics, the physical and biological sciences and medicine. An account of his Jewish origin begins on the second page of *Ex-prodigy* [165]* and continues for a number of pages. He had an interesting theory to account for Jewish devotion to learning. It was in fact the case that a young man who was a good Talmudic scholar, no matter how poverty stricken or unworldly, was considered a good match for the daughter of even the wealthiest family.

---

* The bold-faced numbers in brackets refer to the numbered references in the Bibliography of Norbert Wiener.

Adhering to Orthodox tradition the couple would raise a large family supported by the wife's father or by the wife herself. At the same time over the centuries the learned Christian young man entered the Church and was barred from marriage. This process repeated over tens of generations could have added a genetic bias to the existing cultural bias for learning that prevailed among the Jews.

Leo Wiener was a very intelligent and independent young man. Supporting himself from the age of thirteen onward, he graduated from a Warsaw gymnasium, which was an uncommon achievement for a Russian Jew faced as he was with the anti-Semitic laws of Tsarist Russia and with the insularism of Orthodoxy. There followed an abortive period as a medical student at Warsaw and then an abortive attempt as a technical student at Berlin. All this was over by the time he was eighteen when he emigrated to the United States alone and penniless. He never did get a university education. In the United States he was a factory worker, farmer, peddler, country school teacher, high school teacher, and then a foreign language teacher at the University of Missouri. Always extending his knowledge by intensive reading he eventually became a professor of Slavic languages at Harvard and a philologist of importance.

Norbert's mother, Bertha Kahn Wiener, was born in Missouri. She had one non-Jewish grandparent. However, her family was in the process of being assimilated, all of her mother's brothers having married non-Jews.

Soon after Norbert's birth the family moved to Boston and with his uncommon energy and obvious ability Leo Wiener quickly obtained a position at Harvard. Norbert was a timid anxious child who was sustained by the "solicitude and tenderness," to use his words, of his mother. Bertha Wiener tried to compensate socially for her unconventional husband. Norbert described him as "brilliant," "absent-minded" and "hot tempered." (How well these adjectives applied also to Norbert himself.) Even if this extremely intense man had been a Mayflower descendant, his personality would have made him a maverick at Harvard. Actually, to further aggravate matters, he was an Eastern European Jewish immigrant with no university degree. His unconventionality certainly made his life more difficult. Later when Norbert displayed a similar markedly unconventional personality, his mother tried to smooth over some of the rough spots but to no avail. Had she prevailed, his earlier years would have been much less difficult.

In the summer of 1901 the family travelled to Europe. Norbert particularly remembers visiting Israel Zangwill, the writer and Zionist, and Prince Kropotkin, the Russian nobleman turned Anarchist. (Some thirty years later a student of Wiener's turned out to be the natural son of Kropotkin's secretary.)

Norbert learned to read spontaneously at a very early age and also picked up arithmetic. Already by six he was reading the widest variety of books in his father's library. He was attracted to zoology, physics and chemistry. One article he read as a child excited in him "the desire to devise quasi-living automata" [165, p. 65]. Cybernetics had an early birth!

At seven an attempt was made to enroll him in public school. However he did not fit readily into any grade and his father took him out of school and became his teacher. Anyone who has taught a wife or child, even something as simple as driving a car, knows the tensions that arise in such a family situation. No wonder then that the overly sensitive boy came to regard his perfectionist and highly reactive father an "avenger of the blood" [165, p. 67]. His mother tried to defend him but in his eyes at least was not very successful.

While Norbert realized his father's unusual qualities, nevertheless even forty years later when he became depressed and would reminisce about this period, his eyes would fill with tears as he described his feelings of humiliation as he recited his lessons before his exacting father. Fortunately he also saw his father as a very lovable man and he was aware of how much like his father he himself was.

With his father he went through the mathematics textbooks of Wentworth up to and including analytic geometry. He also learned Latin and German. On his own he read biology texts and a vast amount of literature. The science fiction of H. G. Wells and Jules Verne delighted him. During this period he had many neighborhood playmates and also enjoyed contact with his grandmothers, aunts, uncles and cousins.

When Norbert was nine the family moved to Harvard, Massachusetts and he was enrolled in nearby Ayer High School, which he completed at age eleven. The small town high school was a friendly place for this unusual child and he retained a warm affection for his friends in Ayer.

In 1906, aged eleven, Wiener entered Tufts College. Very wisely his father had decided not to subject the nervous boy to the tension of the Harvard entrance examinations and the publicity that would be given an eleven year old at Harvard. In order to make it unnecessary for Norbert to commute, the family moved to Medford Hillside. His major subject was mathematics but, except for a special reading course in Theory of Equations in his freshman year, the mathematics courses did not go far and the material covered was that suited to train engineering students of that period. The course in Theory of Equations included some Galois theory and was "over my head" [165, p. 104] as Norbert put it.

From his own account of his Tufts education it is clear that physics and chemistry impressed him much more profoundly than his mathematics courses. As we shall see, he was not in fact a prodigy in mathematics. With a neighborhood friend he carried out experiments in electrical engineering. They attempted to execute two of his ideas. Both ideas were excellent and show he was indeed a prodigy in engineering if not in mathematics. His first idea was to design an electromagnetic coherer for wireless messages. It depended on the fact that a magnetic field would compress a mass consisting of iron filings mixed with powdered carbon and thereby decrease the resistance of the mass. The invention of the vacuum tube made this device unnecessary. His second idea was to build an electrostatic transformer by charging a sequence of rotating glass disks through electrodes arranged in parallel and discharging them in series. Years later there was such an apparatus in operation.

He studied some philosophy and read much more of it than was required. Through his father he met William James. In his last year at Tufts his main interest was biology and consequently he was enrolled in 1909, not quite fifteen years old, in the Harvard Graduate School to work in zoology.

He had not been elected to Phi Beta Kappa at Tufts and this hurt him so much that he was opposed to honor societies for the rest of his life. He felt rebuffed and unwanted by the adult community, which he believed was suspicious that prodigies were doomed to failure. At this time he also became acutely aware of the transient character of life and obsessed by fear of death. "Like many other adolescents, I walked in a dark tunnel of which I could not see the issue, nor did I know whether there was any. I did not emerge from this tunnel until I was nearly nineteen years old and had begun my studies at Cambridge University (1913). My depression of the summer of 1909 did not suddenly end; rather it petered out" [165, p. 121].

Unfortunately for Norbert he was a failure at laboratory work which meant that he could not continue in biology. He lacked manipulative skill, good eyesight, and the patience required for meticulous work. It was a hard blow, since he wanted to be a biologist.

There were four other prodigies at Harvard at the time. Among these were A. A. Berle and the composer, Roger Sessions. Both of these men and Wiener had very successful careers. Of the other two, one died early of a ruptured appendix and the other never matured enough emotionally to function successfully.

Norbert's failure in biology led Leo Wiener to urge him to go into philosophy in which he had made a good record at Tufts. Norbert complied but in later years he came to resent what he regarded as an unwarranted interference on his father's part. Actually at the time it was an entirely reasonable proposal. (We shall see that somewhat later it was his father who proposed that he go into mathematics.) The academic year saw Norbert at the Sage School of Philosophy at Cornell. The year was not a success. Problems of adolescence, some inner turmoil about handling his Jewishness, marked social ineptitude and immaturity, and an apparent inability to function well outside of the sheltered family environment caused him to do an average rather than superb job with his courses. He blames vegetarianism, inculcated by his father, for some of his difficulties but this seems far fetched. A fellowship he had been given was not to be renewed. As a consequence his father arranged for his transfer to the philosophy department at Harvard, an action which he again came to resent as an unwarranted interference. On the other hand, with all of his difficulties, his father did regard him as a success.

One of his courses at Cornell had been in the theory of functions of a complex variable but he did not understand it. In part he attributes this to his own immaturity and in part to an overly intuitive approach on the part of the instructor. As to the latter, Osgood's excellent *Lehrbuch* had been published several years earlier and a little enterprise would have turned up this or some other adequate treatment. Hence we can only conclude that, in contrast to his early general intellectual development, in mathematics he was not to be an early bloomer.

The chapter of his autobiography on his Harvard years, 1911–1913, is entitled "A Philosopher Despite Himself." This seems to imply unhappiness or frustration. Actually the tone and content of the chapter indicate that these were two positive and successful years for Norbert. He does let slip at one point that with his sharp mind and keen wit he strove mightily and no doubt with considerable success to be "a thorn in the flesh of my mentors" [165, p. 167]. In so doing he could only have sharpened their awareness of his immaturity.

During this period he studied the border area between philosophy and mathematics with E. V. Huntington of the Harvard Mathematics Department. He refers to Huntington as "a magnificent teacher and a very kind man" [165, p. 167]. Huntington was an expert on setting up systems of postulates which were not redundant and which had nontrivial realizations. Some of Wiener's early research was to be exactly along these lines.

His Ph.D. thesis was in mathematical logic. Owing to the illness of the Harvard philosopher Josiah Royce he worked with Professor Karl Schmidt of Tufts. Schmidt suggested for his topic a comparison between the algebra of relatives of Schroeder and that of Whitehead and Russell. Wiener found the work easy although later, under Russell, "I learned that I had missed almost every issue of true philosophical significance" [165, p. 171]. Subsequently he was to do good independent work in mathematical logic.

The prospect of his written and oral Ph.D. examinations terrified him. In looking back he felt his father helped him very much by distracting him with long walks and by drilling him with numerous questions. By his own standards he feels that he was so paralyzed with fear that he should have been failed in his orals. The examining professors, apparently sensing his fright and knowing his ability from his other work, passed him. He never forgot this experience and was always sympathetic to students showing signs of examination stress.

During his last year at Harvard he applied for and was awarded a travelling fellowship which he decided to use by studying with Bertrand Russell at Cambridge University. He was soon to be nineteen years old. In personality he was at extremes being on the one hand sharply aggressive and on the other immature, dependent and lacking in confidence. He had gone far in his studies but had not yet displayed the creative powers which would make him a great mathematician and then an outstanding philosopher of science.

During the summer of 1913, prior to his departure, Huntington recommended that he read Bôcher's *Modern algebra* and Veblen and Young's *Projective geometry*. Strangely the first book, which was packed with material useful to a budding analyst, did not impress him at the time, although he returned to it many times later. He was still very much the logician and the postulational approach of Veblen and Young appealed to him very much.

The whole Wiener family was to spend the winter in Europe—but he was to be on his own in Cambridge. Russell, the logician and philosopher, was his chief mentor there. Norbert was now mature enough to be on his own and Cambridge was a pleasant social experience. Interestingly enough Russell had a decisive influence in preparing him for his still far away career as a mathematician. Russell suggested that a logician and philosopher of mathematics should take some mathematics courses. Accordingly he took courses with

Hardy, Littlewood and Mercer. Hardy's lucid course covering the essentials of real variable theory including the Lebesgue integral and also an introduction to complex variable theory had a lasting influence on him which however was only to be revealed some six years later.

His main course work was with Russell. The immediate influence of this course on Wiener was profound but in the long run it was of no consequence compared to the effect of certain collateral reading recommendations Russell made to his students. Russell recognized the importance to the philosophy of science of Einstein's theory of relativity and suggested Einstein's three great papers of 1905 as reading material. One of these papers was on Brownian motion a subject in which, within ten years, Wiener was to have one of his greatest mathematical successes. Russell also saw the importance of electron theory and urged Wiener to study it, including the then new work of Nils Bohr. Russell obviously had a very keen sense for what was important in physics.

Many of the Cambridge dons were extreme individualists and markedly eccentric to the point where there was a certain competitiveness to excel in this direction. Wiener, who was presumably conducting himself as decorously as he possibly could, was inherently enough of a deviant so that some of his Cambridge friends thought he had carefully, deliberately and skillfully modified his behavior in order to excel in the competition and accordingly complimented him for his originality. He apparently was now sufficiently sophisticated not to disillusion them.

Wiener's first paper [1], which was on set theory, appeared at this time. His most noteworthy contribution to logic is described objectively by him " . . . in connection with the course [of Russell] I did one little piece of work which I later published; and although it excited neither any particular approval on the part of Russell nor any great interest at the time, the paper which I wrote [4] on the reduction of the theory of relations to the theory of classes has come to occupy a certain modest permanent position in mathematical logic" [165, p. 191]. In all, Wiener was to write some fifteen papers on logic and philosophy before shifting to analysis.

Because Russell was to be away from Cambridge for the May term it was decided that Norbert should finish the year at Göttingen. At Göttingen he again made a good adjustment but was not as strongly influenced by Hilbert, Landau or the philosopher Husserl, as he had been by Russell and Hardy. He appears to have undervalued Landau, who had also been a child prodigy.

Wiener took part in the social life at Göttingen and met many young mathematicians from Europe and the United States. One of these of whom Wiener was very fond was Otto Szasz who Wiener later helped come to the United States when Hitler came to power. Szasz with his gentle modesty, his clear and elegant lectures and his interest, as he put it, in small problems, (likening himself to an artist who painted miniatures), was the very opposite of Wiener.

Wiener submitted an essay [2] to compete for a Bowdoin Prize which Harvard awarded him. He did not regard his paper as having any great merit. During this period he had his first experience with the compulsive obsession and passion that is sometimes necessary for achieving success in research [6]. Of himself he says: "Granted an urge to create, one creates with what one has. With me, the particular assets that I have found useful are a memory of a rather wide scope and great permanence and a free-flowing, kaleidoscope-like train of imagination which more or less by itself gives me a consecutive view of the possibilities of a fairly complicated intellectual situation. The great strain on the memory in mathematical work is for me not so much the retention of a vast mass of fact in the literature as of the simultaneous aspects of the particular problem on which I have been working and of the conversion of my fleeting impressions into something permanent enough to have a place in memory. For I have found that if I have been able to cram all my past ideas of what the problem really involves into a single comprehensive impression, the problem is more than half solved. What remains to be done is very often the casting aside of those aspects of the group of ideas that are not germane to the solution of the problem. This rejection of the irrelevant and purification of the relevant I can do best at moments in which I have a minimum of outside impressions. Very often these moments seem to arise on waking up; but probably this really means that sometime during the night I have undergone the process of deconfusion which is necessary to establish my ideas. I am quite certain that at least a part of this process can take place during what one would ordinarily describe as sleep, and in the form of a dream. It is probably more usual for it to take place in the so-called hypnoidal state in which one is awaiting sleep, and it is closely associated with those hypnagogic images which have some of the sensory solidity of hallucinations but which, unlike hallucinations, may be manipulated more or less at the will of the subject" [165, pp. 212–213].

Cambridge and Göttingen mark the beginning of Wiener's emergence from child prodigy to young scientist. He was old enough to be regarded as a young adult and good enough to be accepted by his peers and elders as at least one of the rank and file of the coming intellectual generation. He was away from his father, an experience needed by most young people but especially important in his case.

Wiener returned to the United States at the beginning of World War I. Harvard again awarded him a travelling fellowship and he again went to Cambridge, which however soon almost ceased to function because of the War. He moved to London. "I looked up another Harvard fellow in philosophy, T. S. Eliot, who I believe, had taken Oxford to himself as I had taken Cambridge to myself. I found him in a Bloomsbury lodging, and we had a not too hilarious Christmas dinner together in one of the larger Lyons restaurants" [165, p. 220]. A short time later he returned to the United States planning to complete his fellowship at Columbia. He did not find the atmosphere there congenial and socially he regressed considerably. His sharp tongue and lack of social sensitivity made him something of a nuisance and accordingly he was treated badly by the graduate students in the dormitory in which he lived.

"My stay in New York also marked my introduction to the American Mathematical Society and my first visual acquaintance with most of the elder scholars of the group. At that time the hotel headquarters was that pile of Gay Nineties respectability, the old Murray Hill Hotel. The Society was then more of a New York institution than it is at the present time, for it had indeed been founded by a New York group, and had been known for some years as the New York Mathematical Society. There attached to it a little of a beer-hall flavor, which has evaporated with time and the increased prosperity and respectability of the scientist" [165, p. 224].

Toward the end of the academic year he was hurriedly summoned home. A former fellow student, who had become an instructor in philosophy, informed the philosophy department that several years earlier Wiener had bribed a janitor to obtain for him some of his actual grades in an examination where the only result legitimately available to the student was whether he had passed or failed. Such an incident had occurred but without any bribery. It did not involve cheating in an examination but was nevertheless unpleasant. Wiener's future career was then being considered by the philosophy department and this charge may well have contributed to his later inability to obtain a position in philosophy. The Harvard philosophy department had earlier promised him the position of assistant and on this they did not renege.

Wiener had joined the Appalachian Mountain Club in 1912 and remained a member for the rest of his life. He was a devoted walker and rock climber. In the summer of 1915, after Columbia, he went on a long trip through the White Mountains with some former class-mates. Then he returned to Harvard to his duties as assistant in philosophy. These included lecturing in elementary philosophy to a section at Harvard and to one at Radcliffe. He also chose to offer a free series of Docent Lectures.

Judging from his own remarks and from his later weakness as an undergraduate teacher his elementary teaching was probably at best passable. In the Docent Lectures "it was my intention to supplement postulational methods by a process according to which the entities of mathematics should be constructions of higher logical type, formed in such a manner that they should automatically have certain desired logical and structural properties" [165, p. 230]. Unfortunately there were some gaps in the reasoning which Wiener had not foreseen but which were picked up by Professor G. D. Birkhoff of the Harvard mathematics department who was attending the lectures. Had these lectures been a success it is plausible that his future in philosophy would have been secure.

Wiener enrolled in Birkhoff's course on the three body problem but could not follow it and discontinued attendance. He was now twenty one. Men of about that age who were never to attain anything like Wiener's ultimate success did follow Birkhoff's courses and indeed in some cases were inspired to the peak research of their careers while still his students. Wiener's coming of age as an analyst apparently could not be hurried. Wiener did become a regular attendant of the Harvard Mathematical Society. There he saw Osgood, Bôcher, G. D. Birkhoff, Huntington and Coolidge. His deep seated insecurity caused him to believe that the Harvard mathematics professors were looking down at the students including himself.

Wiener was living with his family. His parents gave Sunday teas for his father's and his own students. These teas gave him much needed experience in handling himself socially and in making friends. It was at one of these teas that he was later to meet his future wife.

Despite his contact with mathematicians and mathematics, Wiener's research work continued to be mainly in logic, then unfortunately something of a stepchild of both philosophy and mathematics. His work was not regarded as really outstanding even by himself and his productivity fell sharply once he started his assistantship. He was hoping to join a philosophy department but apparently he was not being highly recommended. Given all the facts as they have already been recounted in some of the last half dozen paragraphs, it is not clear how he could have been given high recommendations.

At this point Leo Wiener pressed him to look for a job in mathematics instead of philosophy by registering with teachers' agencies. Thus, nearly twenty two, Wiener turned to a career in mathematics on advice from his father and not spontaneously. We can only guess what his future might have been had some department of philosophy seen fit to hire him in 1916. As it was an instructorship in mathematics turned up at the University of Maine.

The war continued but did not yet involve the United States. During the summer Wiener went to Plattsburg to train for a commission as a reserve officer. Although he was with a group training to become officers and therefore unquestionably somewhat select, he was shocked by their drinking and swearing. Once his tentmates realized how sensitive he was, they taunted him mercilessly. Needless to say he was not judged to be officer material on finishing camp.

The year at the University of Maine was a nightmare. He was inexperienced and almost certainly inept in presenting the elementary courses assigned him. He was younger than his years, over reactive, nervous and an easy victim to baiting. The students made his life miserable. He did manage to meet a few research-minded people led by Raymond Pearl, later a statistician at Johns Hopkins Medical School. "When the war finally came, I asked to be released from my teaching responsibilities to enter some branch of the service, for I was no less eager to leave Maine than the University was to see the last of me" [165, p. 243].

He made some tries at enlisting but his bad eyesight brought rejection. He joined an Officer's Training Corps at Harvard but also to no avail. He spent the rest of the summer reading algebraic number theory and trying his hand at the four-color problem, Fermat's last theorem and the Riemann hypothesis, although his mathematical knowledge at the time was extremely limited. In any case he was apparently becoming more committed to mathematics.

In the fall he obtained a job as an apprentice engineer in the turbine department of the General Electric Co. at Lynn. This was followed by a job for Encyclopedia Americana in Albany, at which he was happy. With the approach of the summer of 1918 he decided to look for a mathematics teaching job again. At that point he was asked to join the ballistic staff at the Aberdeen Proving Ground headed by Professor Oswald Veblen who had been made a major.

At Aberdeen he found Veblen, Bliss, Gronwall, Alexander, Ritt, Bray, Franklin (later his brother-in-law), Gill, Poritzky, Widder, Graustein, and many more. Franklin and Gill, both some four or five years his junior, were his closest friends. About the Proving Ground he writes, "Curiously enough, it furnished a certain equivalent to that cloistered but enthusiastic intellectual life which I had previously experienced at the English Cambridge, but at no American university" [165, p. 258]. He finally managed to enlist as a private but remained assigned at Aberdeen.

With the end of the war came a terrible influenza epidemic. Among its many victims was G. M. Green, a promising young Harvard mathematician, and fiancé of Constance, Norbert's sister. Green's parents gave his mathematics books to Constance, herself a student of mathematics, and in this way Norbert had an opportunity to read them. The books included Osgood's "Funktionen theorie," Lebesgue's book on integration, and Fréchet's book on the theory of functionals, and several books on integral equations. Wiener says, "For the first time, I began to have a really good understanding of modern mathematics" [165, p. 265]. This is an astounding statement from the twenty-four year old Wiener. Five years earlier he had taken Hardy's course, which included the Lebesgue integral and some complex variables. Three years earlier he had been a faithful attendant at the Harvard Mathematical Society meetings. He had tried his hand at the hardest problems in mathematics but apparently had not browsed through the mathematics stacks at the Harvard library. At that time there were relatively few mathematics books and in a few hours one could make a superficial acquaintance with all of them. Had he done so he could easily have purchased such books as that of Osgood, who he of course knew very well personally, of Lebesgue, and of Fréchet. As it was, these books came into his hands purely by chance. One can only conclude that in mathematics, as distinct from logic, his attitude was still very much that of an amateur. Aside from the books of Green he also had access to a set of lecture notes on the Lebesgue integral which his sister Constance had taken at the University of Chicago.

He spent a few months as a reporter for the Boston Herald but did not do well. He then wrote two papers (probably [9] and [10]) on an "extension to ordinary algebra of Sheffer's idea of a set of postulates with a single fundamental operation" [165, p. 269]. Sheffer was at Harvard and had earlier told Wiener about his discovery. Wiener regarded these as "by far the best pieces of mathematical work that I had yet written" [165, p. 269]. However he was soon to leave logic and postulational theory for analysis, the kind of analysis he had learned from the books of Green.

In the spring of 1919, through the good offices of Professor Osgood, Wiener obtained a position for the coming academic year in the mathematics department of the Massachusetts Institute of Technology. At the time this was not a particularly distinguished department but a number of its younger faculty were active in research. The major duty of the department was teaching calculus to engineering students. If the department was not outstanding neither was Wiener at the time. He had never written a paper in what was then regarded as mathematics, he had only an undergraduate degree from Tufts in the subject, his teaching record was at best mediocre, and he was immature and eccentric. One wonders if Osgood would have acted had Leo Wiener not been his friend and colleague. Norbert certainly did not feel indebted to Osgood for finding him the position.

At M.I.T. Wiener's teaching load was more than twenty hours a week of elementary calculus. He was sufficiently energetic so he didn't find this a serious liability. The students were serious and he soon learned how to get along with them. He compensated for his tendency to lecture over their heads by grading liberally. His eccentricities were not regarded as out of place in a college teacher and indeed the students seemed pleased to have at least one teacher who could not be mistaken for anything except a budding college professor.

Since the other members of the mathematics department were not particularly distinguished, the insecure young Wiener did not get the feeling they were looking down at him or that he was under any pressure from them to achieve greatness. Actually M.I.T. was an ideal environment for Wiener at that time when the transition to analysis still lay ahead of him. He felt free to tell his colleagues about his ideas and, if he was unduly and prematurely optimistic about them, they nevertheless encouraged him and did not dampen his ardor. They also tried to buoy him up in his periods of self doubt and deep depression. While they didn't always understand him, they did sense the possibility that his brightness and enthusiasm could lead to real achievement.

In the summer before going to M.I.T. Wiener had a great stroke of good fortune. He received a visit from I. A. Barnett, then himself a young mathematician and a former student of E. H. Moore. Apparently influenced by the book of Fréchet which had been in Green's collection, Wiener aspired to work in functional analysis. He asked Barnett, as one trained in this area, to suggest a good research problem to him. "His reply has had a considerable influence on my later scientific career. He suggested the problem of integration in function space" [165, p. 274]. In fact this suggestion of Barnett completely influenced the whole course of Wiener's work and his greatest achievements all stemmed from this problem. It was a happy accident that

Barnett suggested a problem that turned out to be approachable by the mathematics of the time and yet was sufficiently difficult so that a solution was a real achievement. It was also a problem which when solved was to lead Wiener to more, equally important and pregnant problems.

He worked on this problem for two years before finally solving it. In the course of solving it he studied the Daniell integral and found in the literature unsatisfactory attempts at solution of the problem. An article by G. I. Taylor in the Proceedings of the London Mathematical Society (1920) excited his interest. It was on turbulence. In spirit it was applied rather than pure mathematics. It was important to Wiener for several reasons. One was that he then tried turbulence as a model for his problem. Another reason was that Taylor introduced a type of correlation between a function and its derivative. This was later to suggest to Wiener his autocorrelation and cross correlation functions in generalized harmonic analysis.

The notion of using turbulence as a model for his problem did not work out because turbulence is too complicated. This set Wiener to looking for a simpler model and then he thought of Brownian motion which he had studied in one of Einstein's papers some seven years earlier at Russell's suggestion. The paper of Einstein studied the statistical behavior of particles performing Brownian motion. Wiener's idea was to study the path of a single particle, or rather the ensemble of such paths. Each path became a point in his function space. He was able to show that almost all paths were continuous but non-differentiable. He could compute the average of a functional over the ensemble of paths. He could not have realized at the time the effect his work was to have on the field of probability in the decades ahead. Here is what M. Kac says in 1964: "During the past two decades mathematicians have pursued an even more productive investigation of the theory of 'stochastic processes': the probabilistic analysis of phenomena that vary continuously in time. Stochastic processes arise in physics, astronomy, economics, genetics, ecology, and many other fields of science. The simplest and most celebrated example of a stochastic process is the Brownian motion of a particle.

"The late Norbert Wiener conceived (in 1921) the idea of basing the theory of Brownian motion on a theory of measure in a set of all continuous paths. This idea proved enormously fruitful for probability theory. It breathed new life into old problems . . . . More than that, it opened up entire new areas of research and led to fascinating connections between probability theory and other branches of mathematics" (Sci. Amer. **211** (1964), No. 3, 105).

All of this was to take some years in coming and neither Wiener nor anyone else were aware of the implication of his "Differential space." For Wiener even in the early nineteen twenties the Brownian paths were examples of phenomena which were like turbulence, noise, white light, etc. in that they were not strictly periodic and hence not describable by Fourier series nor were they transient in time and hence were not describable by a classical Fourier integral. He also had the suggestion of Taylor's special case of a correlation function in the study of turbulence. All this was soon to lead him to generalized harmonic analysis. Before this, however, he did several other pieces of work.

In 1920 Wiener had gone to an International Mathematics Congress at Strasbourg. He spent some time with Fréchet whose work he wished to extend. He found a satisfactory set of axioms for vector spaces. But a few weeks later an article by Banach appeared with exactly the same postulates. Wiener's results were published later than Banach's and he added an appropriate reference. Wiener did not stay in this field.

Wiener felt all of his life that Harvard had not treated him well. Actually, as we have seen, Harvard has provided him with two post-doctoral fellowships followed by a year as an assistant in philosophy. Osgood, of the Harvard Mathematics Department, had found the position in the M.I.T. Mathematics Department for him. After Wiener came to M.I.T., he apparently consulted frequently with Kellogg of Harvard about potential theory in which Kellogg was an expert. Kellogg quickly brought him to the research frontiers of this field and Wiener successfully solved some important problems on the behavior of an harmonic function at its boundary which unquestionably brought him more recognition at the time than his Brownian motion work. Unfortunately there arose a problem of timing his publications so as not to compromise the theses of two of Kellogg's students. Actually the matter resolved itself satisfactorily although this is not made clear in Wiener's account. Wiener unfortunately felt that Kellogg had been unfair to him. Thus instead of feeling indebted to Kellogg he again felt that Harvard was mistreating him.

Time has shown Wiener's Brownian motion paper to be more important than his very good work in potential theory. As already stated, Brownian motion is an example of a whole class of physical phenomena the study of which led Wiener to his generalized harmonic analysis. Wiener's own account of how he was led to generalized harmonic analysis is full of references to engineering particularly to communication theory. Most of Wiener's important work was inspired by physics or engineering and in this sense he was very much an applied mathematician. On the other hand, he formulated his theories in the framework of rigorous mathematics and as a consequence his impact on engineering was very much delayed.

"For several years the chief demand made on me at M.I.T. by the electrical-engineering department was to put the Heaviside calculus on a proper logical foundation. Other people were doing the same thing at the same time in other countries although I do not think that any of these treatments were more satisfactory than the one which I ultimately gave. In performing this task, I had to study harmonic analysis on an extremely general basis, and I found out that Heaviside's work could be translated word for word into the language of this generalized harmonic analysis.

"In all this there was an interplay between what I was doing on the Heaviside theory and what I had done on the Brownian motion. Previous to my work there had been no thoroughly satisfactory example given of the sort of motion that would correspond to sound or light with a continuous spectrum—that is, with energy distributed continuously in frequency instead of being lumped in isolated spectrum lines. The harmonic analysis which had already been given corresponded more closely to what one sees when one examines the light of sodium vapor than what one sees when one examines sunlight. (The light of sodium vapor is concentrated in a number of bright lines, whereas sunlight has a continuous distribution in color and consequently in frequency.)" [**177**, p. 78].

"I found that it was possible to generate continuous spectra by means of the Brownian motion or the shot effect and that if a shot-effect generator were allowed to feed into a circuit that could vibrate, the output would be of that continuous character. In other words, I already began to detect a statistical element in the theory of the continuous spectrum and, through that, in communication theory. Now, . . . , communication theory is thoroughly statistical, and this can be traced directly back to my work of that time" [**177**, p. 79].

From G. M. Green's copy of Fréchet's book Wiener was led to ask Barnett for a problem in functional analysis. Barnett's question led Wiener to set up "Differential space" modelled on Brownian motion (studied at the suggestion of Russell) and this in turn led him to generalized harmonic analysis. One of the problems that arose in generalized harmonic analysis led to Wiener's general Tauberian theorems. Such is the role of chance in great discoveries of science. To see how the latter events occur let us look briefly at generalized harmonic analysis.

The class of functions Wiener considered in his harmonic analysis were those $\{f(t)\}$ which were measurable and for which

$$\phi(x) = \lim_{T \to \infty} \frac{1}{2T} \int_{-T}^{T} f(x+t)\bar{f}(t)dt$$

exists for all x. In the special case

$$f(t) = \sum_{1}^{N} a_n e^{i\lambda_n t}$$

λ_n real,

$$\phi(x) = \sum_{1}^{N} |a_n|^2 e^{i\lambda_n x}.$$

If

$$S(u) = \frac{1}{2\pi} \int_{-\infty}^{\infty} \phi(x) \frac{e^{-iux} - 1}{-ix} dx,$$

then $S(u)$ is a monotone step function with jumps of $|a_n|^2$ where $u = \lambda_n$. Thus

$$\phi(x) = \int_{-\infty}^{\infty} e^{iux} dS(u).$$

If u_1 and u_2 are not equal to any λ_n,

$$S(u_2) - S(u_1) = \sum_{u_1 < \lambda_n < u_2} |a_n|^2.$$

Thus $S(u_2) - S(u_1)$ is a measure of the energy of $f(t)$ between frequencies u_1 and u_2. If $f(t)$ is not a trigonometric polynomial, $S(u)$ is still monotone and the above formulas are valid.

From the existence of $\phi(0)$ follows readily that $|f(t)|^2/(1+t^2)$ is integrable $(-\infty, \infty)$ and hence in the sense of convergence in the mean there exists

$$g(u) = \frac{1}{2\pi} \int_{-1}^{1} f(t) \frac{e^{iut} - 1}{it} dt + \frac{1}{2\pi} \left(\int_{-\infty}^{-1} + \int_{1}^{\infty} \right) \frac{f(t)}{it} e^{iut} dt.$$

Plancherel's theorem gives

$$\frac{1}{2\epsilon} \int_{-\infty}^{\infty} |g(u+\epsilon) - g(u-\epsilon)|^2 du = \frac{1}{\pi\epsilon} \int_{-\infty}^{\infty} |f(t)|^2 \frac{\sin^2 \epsilon t}{t^2} dt.$$

To obtain the interesting formula

$$\lim_{\epsilon \to 0} \frac{1}{2\epsilon} \int_{-\infty}^{\infty} |g(u+\epsilon) - g(u-\epsilon)|^2 du = \lim_{T \to \infty} \frac{1}{2T} \int_{-T}^{T} |f(t)|^2 dt,$$

it suffices to prove

$$\lim_{\epsilon \to 0} \frac{1}{\pi\epsilon} \int_{-\infty}^{\infty} |f(t)|^2 \frac{\sin^2 \epsilon t}{t^2} dt = \lim_{T \to \infty} \frac{1}{2T} \int_{-T}^{T} |f(t)|^2 dt$$

in the sense that the existence of either side implies that of the other side. If one sets

$$F(t) = |f(t)|^2 + |f(-t)|^2,$$

the problem becomes

$$(*) \qquad \lim_{\epsilon \to 0} \frac{2}{\pi\epsilon} \int_0^{\infty} F(t) \frac{\sin^2 \epsilon t}{t^2} dt = \lim_{T \to \infty} \frac{1}{T} \int_0^{T} F(t) dt.$$

Wiener attempted to prove (*). He was in Göttingen at the time, as was his friend A. E. Ingham. "It is to Ingham that I owe a scientific lead which has carried me to much of my best work" [177, p. 115]. The word "much" really should be replaced by "some." What Ingham told Wiener when he asked him about (*) was that it was a Tauberian theorem, in the terminology of Hardy and Littlewood, who had solved similar problems. To solve (*) Wiener, instead of trying to adapt the methods of Hardy and Littlewood, devised a highly original and powerful procedure in his general Tauberian theorems. A later lemma used to beautify this theory was his famous theorem on the reciprocal of a nonvanishing absolutely convergent Fourier series.

With $T = e^x$, $\epsilon = e^{-x}$, $t = e^y$ and $F(e^y) = H(y)$, Wiener noticed that (*) becomes

$$\frac{2}{\pi} \lim_{x \to \infty} \int_{-\infty}^{\infty} e^{x-y} \sin^2 (e^{-(x-y)}) H(y) dy = \lim_{x \to \infty} \int_{-\infty}^{x} e^{-(x-y)} H(y) dy.$$

Wiener had recast his problem to the following form: Let $K(x)$ be integrable and satisfy

$$(1) \qquad \int_{-\infty}^{\infty} |K(x)| dx < \infty, \qquad \int_{-\infty}^{\infty} K(x) dx = 1.$$

Let $K_1(x)$ satisfy the same conditions. Then under what conditions does the existence of

(2)
$$\lim_{x \to \infty} \int_{-\infty}^{\infty} K(x-y)f(y)dy = a$$

imply that

(3)
$$\lim_{x \to \infty} \int_{-\infty}^{\infty} K_1(x-y)f(y)dy = a?$$

In the simplest of Wiener's theorems one takes $|f(y)|$ as uniformly bounded. Let $R(x)$ satisfy (1) also. Then from (2) follows readily that

$$\lim_{x \to \infty} \int_{-\infty}^{\infty} R(x-t)dt \int_{-\infty}^{\infty} K(t-y)f(y)dy = a$$

so that if

(4)
$$\int_{-\infty}^{\infty} R(x-t)K(t)dt = K_2(x),$$

then

(5)
$$\lim_{x \to \infty} \int_{-\infty}^{\infty} K_2(x-y)f(y)dy = a.$$

If (5) holds for a class of $K_2(x)$ such as

$$K_2(x) = \frac{\sin^2 Nx}{\pi N x^2}, \qquad 0 < N < \infty,$$

then from the convolution

$$\int_{-\infty}^{\infty} K_1(x-t)dt \int_{-\infty}^{\infty} \frac{\sin^2 N(t-y)}{\pi N(t-y)^2} f(y)dy,$$

one is led to (3). Thus Wiener reduced his problem to the solution of (4) for $R(x)$.

The equation (4) is a convolution. Defining

$$k(u) = \int_{-\infty}^{\infty} K(t)e^{iut}dt$$

etc., one is led to replace (4) by

(6)
$$r(u)k(u) = k_2(u)$$

where

$$k_2(u) = \left(1 - \frac{|u|}{N}\right), \qquad |u| \leq N = 0, \qquad |u| > N.$$

Because

$$(7) \qquad\qquad r(u) = \frac{k_2(u)}{k(u)},$$

Wiener was led to his famous hypothesis

$$k(u) \neq 0, \qquad -\infty < u < \infty.$$

Thus (7) implies

$$(8) \qquad\qquad \int_{-\infty}^{\infty} |r(u)|^2 du = \int_{-N}^{N} |r(u)|^2 du < \infty.$$

In his early papers Wiener added a hypothesis of the type

$$(9) \qquad\qquad \int_{-\infty}^{\infty} |xK(x)|\, dx < \infty$$

which led to the existence of $k'(u)$ as a continuous function and hence
from (7) to

$$(10) \qquad\qquad \int_{-\infty}^{\infty} |r'(u)|^2 du < \infty.$$

From (8) and (10)

$$(11) \qquad\qquad \int_{-\infty}^{\infty} |R(x)|^2 dx < \infty$$

and

$$(12) \qquad\qquad \int_{-\infty}^{\infty} |xR(x)|^2 dx < \infty$$

where

$$R(x) = \frac{1}{2\pi} \int_{-N}^{N} r(u)e^{-iux} du.$$

From (11) and (12)

$$\left(\int_{-\infty}^{\infty} |R(x)|\, dx\right)^2 \leq \int_{-\infty}^{\infty} |R(x)|^2(1 + x^2) dx \int_{-\infty}^{\infty} \frac{dx}{1 + x^2} < \infty$$

and so $R(x)$ satisfies (1) and the simplest of Wiener's Tauberian theorems is proved.

The condition (9) is satisfied in all cases of interest, but Wiener was unhappy with it and eventually (1931) replaced it by his famous theorem that if

$$g(x) = \sum_{-\infty}^{\infty} a_n e^{inx}$$

where

$$\sum_{-\infty}^{\infty} |a_n| < \infty$$

and if $g(x) \neq 0$, then $1/g(x)$ has an absolutely convergent Fourier series. This latter result was later recast by Gelfand in an abstract form which is probably much better known in contemporary mathematics than the actual Tauberian theorems of Wiener.

Wiener's first work in generalized harmonic analysis appeared early in 1926 and his Tauberian theorems in 1928. In 1924, Wiener had been promoted to an assistant professorship at M.I.T. But Wiener was not satisfied with his rate of promotion or with the fact that he was not invited to join one of the several departments with more prestige than that at M.I.T. In part he was merely paying the usual price, not a large one, for being an innovator. He was to receive ample recognition before long but he was extremely impatient. He resented what he thought was mistreatment by established mathematicians particularly those at Harvard. " . . . I knew very well that I was competitive beyond the run of younger mathematicians, and I knew equally that this was not a pretty attitude. However, it was not an attitude which I was free to assume or to reject. I was quite aware that I was an out among ins and that I would get no shred of recognition that I did not force. If I was not to be welcomed, well then, let me be too dangerous to be ignored" [177, p. 87]. This was hardly an attitude conductive to winning good will. Fortunately it is an exaggerated and lopsided account. Actually he could also be generous and jovial even though he was plagued by something of a persecution complex.

Because of their own mathematical interests J. D. Tamarkin of Brown, an émigré from the Soviet Union, and G. H. Hardy were among the first to appreciate Wiener's generalized harmonic analysis and Tauberian theorems and they helped establish his reputation. "He [Hardy] also thought well of my work, and between him and

Tamarkin I began to be heard of in this country, but I was never able to forget that the people to whom I owed the greatest part of my recognition were not Americans" [177, p. 130]. In fact, with much less delay than often occurs in science, Wiener was to be accorded ample recognition from the American mathematical community.

In 1926 Wiener married Margaret Engemann, a young woman of considerable character and ability, who had emigrated with her widowed mother and brothers from Germany at age fourteen. His parents apparently seeing in her the maturity Norbert required in a mate, and probably anxious to get their aging fledgling out of the nest, strongly approved of the match. As Norbert's wife she would require tact, forbearance, and a willingness to assume almost all family responsibility, to shelter Norbert from outside distractions, to humor him when depressed, to allay his fears and anxiety, and to tolerate him in his unbounded flights of fancy when he was cheerful. She could not have known the job she was undertaking but she proved equal to it. She gave him as much peace of mind as was humanly possible so that he could pursue his work.

Norbert had become attached to the town of Sandwich in New Hampshire. His parents helped the young couple buy a home there to which they were to go for their vacations from then on. The house was set in a valley with the Ossipee Mountains to the south and Chocorua towering in the north. Kline, Wintner and other mathematicians were to come there regularly as a result of visiting the Wieners.

In 1929 Wiener was promoted to an associate professorship at M.I.T. His long paper on generalized harmonic analysis appeared in Acta Mathematica in 1930. Since he was not a good expositor Tamarkin helped with the preparation of this paper. There followed an equally important memoir on Tauberian theorems that appeared in the Annals of Mathematics early in 1932. With the appearance of these two important papers he was promoted to a professorship in 1932.

The year 1931–1932 was spent by the Wieners and their two small daughters, Barbara and Peggy in Cambridge, England. Here, on Hardy's invitation, Wiener gave a series of lectures which he then prepared as a book, *The Fourier integral* [81], published by the Cambridge University Press. Aside from presenting his work on absolutely convergent series, Tauberian theorems and generalized harmonic analysis, Wiener gave a very interesting proof of the Plancherel theorem based on his observation that the Fourier transform can be regarded as an operator having as its eigenfunctions the Hermite orthogonal functions with eigenvalues that are 1, i, -1, or $-i$.

Before he had left for Cambridge, Wiener had collaborated with E. Hopf on the solution of the integral equation

$$f(x) = \int_0^\infty K(x - y)f(y)dy$$

for $f(x)$. This equation occurs commonly in many applications and has come to be known as the Wiener-Hopf equation. In its solution a study of the Fourier transform in the complex plane is necessary. The Fourier transform as a function of a complex variable proved to be very important in Wiener's subsequent research.

In the year 1932–1933, back at M.I.T., Wiener had the young British mathematician R. E. A. C. Paley as a visitor. They collaborated on a number of theorems involving the Fourier transform in the complex plane. The result of their collaboration was the book *Fourier transforms in the complex domain* [92], an American Mathematical Society Colloquium Publication, 1934. Paley was killed in a skiing accident almost a year before the book was completed and most of the actual writing of the book was done by Wiener. Wiener was extremely generous in matters relating to collaboration and with no hesitation he listed Paley as a coauthor of the book. The most widely known theorems in the book have turned out to be those appearing in the introduction. One of these reads as follows: Let $F(s)$ be analytic for $-\lambda \leq \sigma \leq \mu$, $s = \sigma + it$, and let

$$\int_{-\infty}^\infty |F(\sigma + it)|^2 dt \leq \text{const}, \qquad -\lambda \leq \sigma \leq \mu.$$

Then there exists a measurable $f(x)$ such that

$$\int_{-\infty}^\infty |f(x)|^2 e^{2\mu x} dx < \infty,$$

$$\int_{-\infty}^\infty |f(x)|^2 e^{-2\lambda x} dx < \infty,$$

and, for $-\lambda \leq \sigma \leq \mu$,

$$F(\sigma + it) = \lim_{A \to \infty} (2\pi)^{-1/2} \int_{-A}^A f(x) e^{x(\sigma + it)} dx.$$

(Actually it suffices for $-\lambda < \sigma < \mu$ and the case $\lambda = 0$, $\mu = \infty$ is so formulated by them.)

It is interesting that Wiener who, as an innovator resented very much the more rapid recognition accorded the young scientist working in already accepted areas, should on achieving success himself, so strongly approve of those doing work in established areas close to his own as was the case with Paley and myself. Actually it seems inevitable that this be so and hence that the innovator has to expect the possibility of some delay in being recognized.

In 1933 Wiener was honored by the American Mathematical Society with the award of the Bôcher prize which he shared with Marston Morse. The prize is awarded every five years, and since Wiener's memoirs on generalized harmonic analysis and Tauberian theorems had appeared in 1930 and 1932 respectively, it is evident that the Society accorded him its highest honor as soon as it was possible to do so. At about the same time he was elected to the National Academy of Sciences. He stated that the crass bargaining for votes that preceded the annual elections of new members disgusted him and he quickly resigned.

In the summer of 1934 the American Mathematical Society again honored Wiener by inviting him to present the Colloquium Lectures. He presented material from his book with Paley at a pleasant meeting in Williamstown, Massachusetts. A little later he was elected a Vice President of the American Mathematical Society. His dislike for administrative duties precluded his holding the Presidency. In 1949 Wiener was invited to give the Willard Gibbs Lecture at the Annual Meeting of the Society.

Wiener had made frequent trips to Europe through 1927. In fact in 1926–1927 he had been a Guggenheim Fellow in Europe. As already stated 1931–1932 was spent in Cambridge, England. In 1935–1936 he was a visiting professor in China. In 1947, 1949, and 1951 he spent the fall terms in Mexico at the National Institute of Cardiology. In 1951 he was also a Fulbright Lecturer in Paris. In 1955–1956 he was a visiting professor in Calcutta, India. In 1960 and 1962 he spent a term at the University of Naples and in 1964 at the time of his death occupied a visiting position in Holland.

Wiener had an international outlook. Not only did he feel at home in Western Europe but also in Mexico, China and India. While he had visited the Soviet Union only briefly, he had long admired their achievement in mathematics.

I became acquainted with Wiener in September 1933, while still a student of electrical engineering, when I enrolled in his graduate course. It was at that time really a seminar course. At that level he was a most stimulating teacher. He would actually carry on his research at the blackboard. As soon as I displayed a slight comprehension of what he was doing, he handed me the manuscript of Paley-Wiener for revision. I found a gap in a proof and proved a lemma to set it right. Wiener thereupon sat down at his typewriter, typed my lemma, affixed my name and sent it off to a journal. A prominent professor does not often act as secretary for a young student. He convinced me to change my course from electrical engineering to mathematics. He then went to visit my parents, unschooled immigrant working people living in a run-down ghetto community, to assure them about my future in mathematics. He came to see them a number of times during the next five years to reassure them until he finally found a permanent position for me. (In those depression years positions were very scarce.) If this picture of extreme kindness and generosity seems at odds with Wiener's behavior on other occasions, it is because Wiener was capable of childlike egocentric immaturity on the one hand and extreme idealism and generosity on the other. Similarly his mood could shift quickly from a state of euphoria to the depths of dark despair.

For several years after the publication of the Paley-Wiener book Wiener's mathematical work consisted mainly of extensions of his earlier theorems and methods. Much of this work was very good but was probably not his very best. In his paper, *The homogeneous chaos*, (1938) [108], an important added ingredient is the ergodic theorem. The addition of the ergodic theorem to his earlier techniques is probably the main feature that distinguishes Wiener's later mathematical work from that done in 1921–1934. On the advice of Hopf and Jessen, Wiener had already used the ergodic theorem in the last chapter of the Paley-Wiener book [92] but his really deep interest in it developed later.

During this period Wiener made the acquaintance of a Mexican physiologist Arturo Rosenblueth, at that time a collaborator of Walter Cannon at the Harvard Medical School. There began a collaboration which continued in Mexico after the War when Wiener spent several terms as a visitor of the National Institute of Cardiology. This collaboration certainly played a part in the development by Wiener of cybernetics.

Wiener felt uneasy about his mathematical work during the years immediately preceding World War II and pressed his colleagues to affirm that his productivity was not declining. He had always needed approval from those around him. His usual words of greeting became, "Tell me, am I slipping?" Whether one knew what he had been doing or not the only response anyone ever made was a strong denial. However this was usually not enough and it was necessary to affirm in the strongest terms the great excellence of whatever piece of his research he himself would proceed to describe sometimes in the most glowing terms. Altogether such an encounter was an exhausting experience. Indeed at the time there was an instructor at M.I.T. slightly younger than myself who would complain bitterly about the expense of a chance meeting with Wiener. The man, now a very successful mathematician and known for his exuberance and bounce, found that the degree of enthusiasm he had to feign to meet the demands of Wiener left him enervated and without confidence in his own work. Such was his state that he would rush off to a psychiatrist. The fee of five dollars (25 years ago) was much more than he could afford. Whether this played a role in his early departure from M.I.T. he never said.

In the years preceding the War, Wiener became a supporter of the Spanish Loyalists. As a recent visitor to China he was also moved to be active in the political support of the Chinese in their struggle against the Japanese invasion of their country. Several of his younger friends and colleagues had become Communists—indeed Stalinists. Wiener saw in Stalinism a dogmatic new religion with naively simple formulas for curing the ailments that afflict humanity and he rejected it. Instead he reacted as an idealist and humanitarian to the rising turmoil of a world speeding toward disaster. He succeeded in helping a few colleagues, some Jewish and some Socialists, flee from Hitler Germany to the United States. It was during this period that he had his first experiences on the public lecture platform and he was good at it.

When the War finally came to Europe the United States began to mobilize its scientists. Wiener obtained a small grant to work on fire-control. The problem was to design apparatus that would direct anti-aircraft guns effectively. With the increase in speed of planes this was a difficult problem. It was necessary to determine the position and direction of flight of the aeroplane and then extrapolate over the flight time of the projectile to determine where the plane would be so that the projectile could be aimed so as to reach it. This problem stimulated Wiener to produce his prediction theory and also eventually brought him to cybernetics.

The mathematical problem of prediction as he formulated it was solvable by a synthesis of his own previous work. He could have handled it readily any time after 1931, had he conceived of the problem. As one well versed in the theory of linear electrical circuits and of the mathematical tractability of least squares procedures,

Wiener formulated his problem more or less as follows: Consider a function of time $f(t)$ which is the sum of a function $g(t)$, (which could be a coordinate of the moving aeroplane) and a noise $f(t) - g(t)$. How best determine $g(t+h)$ for some $h > 0$ from a knowledge of $f(t-\tau)$ for $\tau \leq 0$? Method of solution: Choose $K(\tau)$, $\tau \geq 0$, so that

$$\lim_{T \to \infty} \frac{1}{2T} \int_{-T}^{T} \left[g(t + h) - \int_{0}^{\infty} K(t - \tau) f(\tau) d\tau \right]^2 dt$$

is minimized.

Proceeding informally the standard variational technique leads to an integral equation of the first kind for K:

$$(13) \qquad \int_{0}^{\infty} \phi(t - \tau) K(\tau) d\tau = \chi(t + h), \qquad t \geq 0.$$

Here ϕ is Wiener's autocorrelation function for f,

$$\phi(t) = \lim_{T \to \infty} \frac{1}{2T} \int_{-T}^{T} f(t + \tau) f(\tau) d\tau,$$

and

$$\chi(t) = \lim_{T \to \infty} \frac{1}{2T} \int_{-T}^{T} g(t + \tau) f(\tau) d\tau,$$

Wiener's cross correlation function for g and f.

The equation (13) is somewhat reminiscent of the Wiener-Hopf equation. It can be treated by the techniques available from the theory of Fourier transform in the complex domain. Actually Wiener's treatment of it was not the simplest possible. Also as in previous cases where Wiener started from a physical problem, his exposition was not directed to the engineer but rather to the pure mathematician. In earlier cases this had delayed the practical utilization of his discoveries, but there were trained mathematicians in the war laboratories who were able to adapt his ideas to the needs of applications. In some cases the modifications of his ideas for practical applications so trivialized the mathematics as to actually cause him pain. On the other hand he was gratified that his work was being used even where in one case as finally simplified, it boiled down to a slightly modified form of Gauss' least squares procedure and led to nothing more sophisticated mathematically than a system of n linear algebraic equations in n unknowns.

Wiener's prediction theory did have an important influence in engineering. His introduction of the statistical element in the question of the existence and significance of the auto- and cross-correlation functions proved very important, and his generalized harmonic analysis finally became a tool of applied mathematics. From the strictly mathematical point of view this appears to be the main contribution of Wiener to cybernetics which we will now consider.

In fire control as first examined in 1940–1941 by Wiener, the role of the human operator, who turned his cranks so as to keep a moving target on the cross hairs of his telescope, brought human neurophysiology into this problem. Wiener began to consider the reactions of the human being in terms of the theories of the communications engineer and found that feedback and stability which were of great importance in control problems in engineering were equally important in neurophysiology. We recall that Wiener had wanted very much to be a biologist but had failed in his first year of graduate school because he lacked the ability to do satisfactory laboratory work. Some thirty five years later he was able to make his mark in biology by way of his theory of control and communication as applied to man and machine in the theory he called cybernetics.

A part of the scientific material that enters into Wiener's cybernetics is from the work of others. His role in cybernetics was not only that of an innovator but also that of a publicist, synthesizer, unifier, popularizer, prophet and philosopher. He has been called the philosophizer of the age of automation. In cybernetics he returned to the field of philosophy he had left some twenty five years earlier.

"The whole background of my ideas on cybernetics lies in the record of my earlier work. Because I was interested in the theory of communication, I was forced to consider the theory of information and, above all, that partial information which our knowledge of one part of a system gives us of the rest of it. Because I had studied harmonic analysis and had been aware that the problem of continuous spectra drives us back on the consideration of functions and curves too irregular to belong to the classical repertory of analysis, I formed a new respect for the irregular and a new concept of the essential irregularity of the universe. Because I had worked in the closest possible way with physicists and engineers, I knew that our data can never be precise. Because I had some contact with the complicated mechanism of the nervous system, I knew that the world about us is accessible only through a nervous system, and that our information concerning it is confined to what limited information the nervous system can transmit.

"It is no coincidence that my first childish essay into philosophy, written when I was in high school and not yet eleven years old, was called *The theory of ignorance.* Even at that time I was struck with the impossibility of originating a perfectly tight theory with the aid of so loose a mechanism as the human mind. And when I studied with Bertrand Russell, I could not bring myself to believe in the existence of a closed set of postulates for all logic, leaving no room for any arbitrariness in the system defined by them. Here, without the justification of their superb technique, I foresaw something of the critique of Russell which was later to be carried out by Gödel and his followers, who have given real grounds for the denial of the existence of any single closed logic following in a closed and rigid way from a body of stated rules.

"To me, logic and learning and all mental activity have always been incomprehensible as a complete and closed picture and have been understandable only as a process by which man puts himself *en rapport* with his environment. It is the battle for learning which is significant, and not the victory. Every victory that is absolute is followed at once by the Twilight of the gods, in which the very concept of victory is dissolved in the moment of its attainment.

"We are swimming upstream against a great torrent of disorganization, which tends to reduce everything to the heat-death of equilibrium and sameness described in the second law of thermodynamics. What Maxwell, Boltzmann, and Gibbs meant by this heat-death in physics has a counterpart in the ethics of Kierkegaard, who pointed out that we live in a chaotic moral universe. In this, our main obligation is to establish arbitrary enclaves of order and system. These enclaves will not remain there indefinitely by any momentum of their own after we have once established them. Like the Red Queen, we cannot stay where we are without running as fast as we can.

"We are not fighting for a definitive victory in the indefinite future. It is the greatest possible victory to be, to continue to be, and to have been. No defeat can deprive us of the success of having existed for some moment of time in a universe that seems indifferent to us.

"This is no defeatism, it is rather a sense of tragedy in a world in which necessity is represented by an inevitable disappearance of differentiation. The declaration of our own nature and the attempt to build up an enclave of organization in the face of nature's overwhelming tendency to disorder is an insolence against the gods and the iron necessity that they impose. Here lies tragedy, but here lies glory too.

"These were the ideas I wished to synthesize in my book on cybernetics" [177, pp. 323–325].

Wiener's entry into cybernetics did not represent any sudden shift on his part. I have already mentioned several times the importance of physical problems in motivating Wiener. Wiener was well aware of this and often discussed the physical origin of much of his work. However his work such as generalized harmonic analysis and of course that on Tauberian theorems, to which his harmonic analysis had led him, were so rigorously formulated as pure mathematical theories that there was some skepticism about the source of his inspiration. G. H. Hardy once asked me whether Wiener's claims about the applied origin of his work was not a "pose." In my opinion it was decidedly not a pose. In fact a perusal of *I am a mathematician* reveals a large number of references to physicists including Gibbs, Maxwell, Planck, Einstein and Born, and to Heaviside. Wiener's comments reveal an intimate acquaintance with and a deep appreciation for their work. He also has references to a number of mathematicians but says almost nothing about what they did with the exception of Lebesgue whose work is described in almost physical terms. We saw that although he knew and admired Hardy it was Ingham who told him about the work of Hardy and Littlewood on Tauberian theorems in 1926. Indeed Wiener was largely unfamiliar with the mathematical work of some of the mathematicians he admired very much.

One of his several major aspirations incidentally was to be as important a contributor to electrical engineering as Heaviside was. I believe he succeeded in this aim. Wiener was a great admirer of Heaviside. He wrote a novel, *The tempter* [195]. Wiener informed me that his inspiration for the character Woodbury was Heaviside while Dominguez was suggested to him by Pupin.

Wiener's later work in mathematics was done almost entirely in collaboration with other mathematicians. Perhaps the best of this work is that which arose from the problem of extending prediction theory to several variables. In my opinion most of his work in cybernetics was not mathematical.

Through cybernetics Wiener had become widely known all over the world to people who regarded his mathematics as rather dry, narrow, incomprehensible technical work. To a considerable extent he lost contact with the younger members of the mathematical community. In the preface of his book *Ex-prodigy* [165] which appeared in 1952 he mentions sixteen people to whom he had submitted the manuscript for criticism. Not one of these is a mathematician. In his last years he probably knew none of the mathematicians under 35 in the Greater Boston Area with the possible exception of Rota with whom he could converse in Italian. Wiener was proficient in most of the languages of Western Europe and also had a limited conversational ability in Chinese. He loved to use these languages.

With the publication of his book *Cybernetics* [138] in 1948, Wiener had quickly become a public figure. Over the years an International Association for Cybernetics has been formed, international congresses are held, and an international journal exists. In the Soviet Union where Wiener was first accused of propounding bourgeois heresy, a later liberalization saw the formation of Institutes of Cybernetics.

It is of interest to see Wiener as he appeared to those who knew him only as the founder of cybernetics. The following excerpts are from an article by S. Toulmin, Director of the Nuffield Foundation Unit for the History of Ideas in London, which appeared as the lead article in the New York Review **3**, No. 3, Sept. 24, 1964.

"He was (to use the word in an entirely innocent sense) the most *peculiar* American in my experience, and even in England I can liken him only to the late Sir Thomas Beecham. The similarities between the two men were no accident. True: they had a certain physical resemblance. Both of them were short, myopic, tubby. But their rotundity was more than a genetic coincidence: in both men it marked out the cosmopolitan, the *bon vivant*. [In fact he was a vegetarian and his moral outlook was Puritanical.] With it, there went a rotundity of expression in public conversation—I nearly said monologue— which was too puckish to be called pompous, and an assumed air of prejudice and self-importance so extreme it became a joy to observe. For so many of the barbs in which both men indulged were penetrating and well-placed; and after all, on these occasions their tongues were never far from their cheeks . . . "

" . . . he graduated Ph.D. from Harvard at eighteen. After seven more years of study and teaching in Europe and the U. S., he found his life's niche at the Massachusetts Institute of Technology and there he took root, gradually building up a solid though unspectacular reputation with papers on a variety of mathematical subjects—notably on Fourier integrals, about which he published a book in 1933."

"The result [of his war work] was 'cybernetics'—i.e., the general theory of control and communication—and it was in this field that Wiener was to spend the second, and more striking half of his professional career."

"If the idea of a 'thinking machine' was to escape from being a contradiction in terms, . . . it was necessary to develop a systematic mathematical theory, capable of representing, without distinction, both the interactions of brain and limbs involved in rational behavior, and also the networks and linkages required in order to construct an artefact capable of simulating that behavior; . . . (After all, the power of the 'new philosophy' lay not in bare appeals to mechanical analogies, but in its method of proof by way of mathematics . . . .) . . . the . . . fundamental mathematical victory is owing above all to Norbert Wiener."

This is how Wiener's cybernetics appears to a nonmathematician, who regards cybernetics as a mathematically demonstrated theory. In fact, cybernetics is so broad that it probably cannot be viewed as a mainly mathematical theory capable of being demonstrated in the spirit of mathematics. The more sharply defined and narrower but correspondingly deeper areas associated with Automata Theory or Information Theory are certainly branches of applied mathematics, but it is not clear that cybernetics can be so classified.

Not many weeks before his death on March 18, 1964 Wiener was invited to Washington to receive from President Johnson The National Medal of Science. It was a fine culmination to a career devoted to the goal of achieving excellence in creative scientific work. Wiener was surprised when he learned that among the organizations that had proposed his name for this distinction was the American Mathematical Society. He was very impressed with the appropriateness of the citation accompanying the award and was almost unable to believe that it had been prepared from a draft submitted by a Princeton mathematician. To the end he could not realize that he was really accepted by the American mathematical community.

Massachusetts Institute of Technology

FROM PHILOSOPHY TO MATHEMATICS TO BIOLOGY[1]

BY WALTER ROSENBLITH AND JEROME WIESNER

Two months before his death, in a ceremony at the White House, Norbert Wiener was awarded the National Medal of Science. The citation by President Johnson said: " . . . for marvelously versatile contributions, profoundly original, ranging within pure and applied mathematics, and penetrating boldly into the engineering and biological sciences."

Our assignment here is twofold: we want to explore how Wiener came to penetrate into biology, a field into which few real mathematicians had strayed before him; we should also like to assess, no matter how incompletely, the imprint that Wiener has left upon the sciences of Life and Man.

From his early youth Wiener, the prodigy, acquired intensive experience in the manipulation of both mathematical and linguistic symbols; but his career choice seemed initially little related to these skills. Perhaps in part due to his father's acquaintance with Walter B. Cannon, Norbert seemed sufficiently interested in biology to become a graduate student in zoology at Harvard University, after he had graduated at the age of 14 from Tufts College. But, in spite of his interest in the subject matter, Norbert had neither the manual skill nor the patience to do well in the graduate courses in biology of that era. In one of his autobiographical books Wiener commented on the contrast between his quick insight into ideas and his extreme lack of manual dexterity as follows:

"This impatience was largely the result of a combination of my mental quickness and physical slowness. I would see the end to be accomplished long before I could labor through the manipulative stages that were to bring me there. When scientific work consists in meticulously careful and precise manipulation which is always to be accompanied by a neat record of progress, both written and graphical, impatience is a real handicap. How much of a handicap this syndrome of clumsiness was I could not know until I had tried. I had moved into biology, not because it corresponded with what I knew I could do, but because it corresponded with what I wanted to do.

"It was inevitable that those about me discouraged me from further work in zoology and all other sciences of experiment and ob-

---

[1] Reprinted with permission of the authors and publisher from *The Journal of Nervous and Mental Disease*, Vol. 140, No. 1, Copyright © 1965, The Williams & Wilkins Company.

servation. Nevertheless, I have subsequently done effective work together with physiologists and other laboratory scientists who are better experimenters than I, and I have made some definite contributions to modern physiological work."

After a short period as a "philosopher despite himself," Wiener found his way into mathematics via a doctoral dissertation in the area of Russellian logic, and a few "Wander"-semesters at Cambridge (on the Cam) and Göttingen. The involvement of America in World War I brought Norbert Wiener to the Aberdeen Proving Grounds and involved him in the computation of ballistic tables. After a short and not too happy interlude as a journalist, Wiener joined the Massachusetts Institute of Technology Mathematics Department in 1919. Although during the next 45 years Wiener remained a productive member of this Department, he had an important influence in many other departments of the Institute as well. Few are the fields of science, engineering, social science or even humanistic scholarship which Wiener's thoughts failed to stir up, often in a rather unorthodox manner. Wiener's presence at M.I.T. spans the period during which the Institute transformed itself from a technical school into a university of a novel type, one "polarized around science," and his intellectual virtuosity, curiosity and integrity contributed importantly to that transition.

When Wiener came to M.I.T., the Mathematics Department was predominantly a service department concerned with preparing students for engineering careers. In a manner which more pure mathematicians in the United States could emulate, Wiener did not hesitate to become interested in the problems of his engineering colleagues. When many years later the great English mathematician, Hardy, claimed that Wiener's engineering terminology was mere camouflage, he misunderstood both Wiener's motivations and sense of social responsibility. Even the purest of mathematics can be a potent tool in very practical pursuits, and Wiener felt that mathematicians, to be effective, need to realize that their labors are changing the nature of society.

Most of Wiener's later mathematical work stemmed from his early interest in the study of irregularities and in his attempts to give meaningful mathematical descriptions of such irregularities, no matter where in nature they occur. His study of Brownian motion led him to study forms of harmonic analysis more general than the classical Fourier series and the Fourier integral. He developed both auto- and cross-correlation analysis and related them to the established forms of spectral analysis.

Under Karl T. Compton's presidency of M.I.T., the Departments of Physics, Chemistry and Mathematics ceased being merely service departments and became the nucleus of a School of Science with increased commitment to what is today called Basic Research. But this enhanced status of mathematicians and of mathematical research did not lead Wiener to loosen his ties with his colleagues in the School of Engineering. On the contrary, during the 1930s, Vannevar Bush and several younger faculty members from M.I.T.'s Electrical Engineering Department interacted with Wiener in ways that came to affect the future of computer and communications engineering significantly, particularly through the use of sophisticated mathematical techniques.

Bush was beginning to overcome the technical obstacles that stood in the way of the construction of the differential analyzer, the pre-World War II forerunner of modern high-speed computing machinery. Wiener's close contact with this program and his joint work with Y. W. Lee on the design of electric circuits, led him to consider, most often in the abstract, the potentialities of the computers of the future and to search for criteria and concepts that would make it possible to separate message from adventitious noise. This was also the period during which Arturo Rosenblueth[2] and Norbert Wiener examined closely—in a series of monthly discussion meetings—how the scientific method was being applied in a variety of fields.

During World War II Norbert Wiener worked on the design of fire control apparatus for anti-aircraft guns. In some sense this problem seemed to be tailor-made for him since it permitted him to pull together many of his previous interests: Wiener saw with great clarity the close relation between the statistical study of time series and the formulation of the basic task of communication engineering, namely the transmission of messages. For a message to be transmitted there must be a repertory or ensemble of possible messages and a way of assessing the probability of these messages. These topics when joined to the problems of filtering and prediction from existing time series comprise the substance of a book that Wiener wrote under the title of *Extrapolation, interpolation and smoothing of stationary time series (with engineering applications)*. (During the war it was known by the more picturesque title of the "Yellow Peril"—the cover contributed the adjective.)

The study of fire-control problems led Wiener, for the first time, to

---

[2] Arturo Rosenblueth, at that time at Harvard Medical School, was the brilliant associate of Walter B. Cannon whose statement of the body's regulatory and stabilizing mechanisms represents one of the high points of classical physiology.

deal directly with man in his coupling to a machine. Human sensory and motor abilities were involved in tracking behavior. The way man corrects for errors led to a general consideration of feedback mechanisms in stabilizing performance and of the particular pathology known as "hunting" in the presence of overapplication of feedback.

The type of analysis Wiener and his co-workers used was not too different in inspiration from the homeostatic, error-correcting mechanisms Cannon had dealt with, both conceptually and experimentally. Wiener and his engineering associate, Bigelow, therefore quite naturally consulted with Arturo Rosenblueth on "intention tremors" as the human pathological symptom that was most closely akin to the pathology encountered in servomechanisms. From this collaboration resulted the famous manifesto, "Behavior, Purpose and Teleology" [127].[3]

Wiener emerged from this wartime work with a conviction that communications engineering, the behavior of servomechanisms, of computing machines and of the nervous system, could all be regarded from a unified overall viewpoint. Drawn by his enthusiasm, people from these disciplines and fields gathered around Wiener in Princeton in a meeting that foreshadowed the famous Macy Foundation series of conferences on Cybernetics.

These conferences, which were the prototype of the many multidisciplinary symposia that followed in the next two decades, subjected the fields that came to be known as sciences of communication to a searching examination. To Wiener and many of his colleagues, communication was clearly the cement of the nervous system, of society, of any complexly organized structure. There was less than unanimity among them and the scientific community at large as to whether cybernetics was a unifying science, a common basis for thought, a convenient common language for functionally related problems, a set of analogies or a program. Many sensational, science-fictionish tales have been written and told under the cybernetic label (not by Wiener who when writing science fiction stories labeled them accordingly) but often by people of whom he was not as critical as he was of his closer colleagues.

The question has often been asked what specifically has Wiener (or cybernetics) contributed to fields other than mathematics. In the area of engineering there can be no doubt that his lifelong association with colleagues and problems has left a profound mark upon the way

---

[3] The bold-faced numbers in brackets refer to the numbered references in the Bibliography of Norbert Wiener.

much of engineering is practiced, conceptualized and taught. When it comes to the biological sciences, it is much harder to identify discrete contributions that he has made or problems that he has "solved." Since neither of us was trained in this area of science, we feel compelled to be cautious. But it seems that the work Wiener did (mainly with Rosenblueth) and especially the work he *stimulated* on such topics as biological regulation, characterization of the electroencephalogram as a time series, prosthetic devices, and much else has been of real substance and significance.

We had the good fortune to belong to the post-World War II Wiener Supper Seminar, we were his colleagues on the M.I.T. faculty, and saw at first hand his influence pervade the Institute and, in particular, the Research Laboratory of Electronics. We can offer personal testimony to the quasi-generality of his own remarks on his cooperation with the outstanding British mathematician Paley [92]; he wrote, "My role was primarily that of suggesting problems and the broad lines on which they might be attacked, and it was generally left to Paley to draw the strings tight." We, too, felt enriched by the problems he suggested and by the excitement he generated by his views of what needed to be done.

In areas in which Wiener's intuition was less educated than in engineering, he was often impatient with experimental details; for example, he seemed sometimes unwilling to learn that the brain did not behave the way he expected it to. Yet even those who have been most critical of his interpretation of data and his speculations can not fail to pay homage to the seminal influence he has exercised by acquainting many a scientist with a different kind of potentially relevant concepts and mathematical and engineering techniques. He had relatively little contact with the spectacular progress in molecular biology which took place after *Cybernetics* [138] had been written. This is a pity; for these challenging problems of the flow of intra- and inter-cellular information might have benefitted from an attack by his fertile intellect. He might have found here a still deeper understanding of the logical processes that are at work in living systems, whose nature so intrigued him.

Wiener solved neither all the problems of the brain nor those of the communication sciences; but he articulated them and gave many scientists a new way of looking at some of the most puzzling questions.

Our country was indeed fortunate to have had the complementary influences of Wiener, Von Neumann and Shannon at work in the immediate postwar period when our educational institutions went through a great expansion in research and graduate study. It was under the auspices of this *Zeitgeist* that young people from such different fields as neurophysiology, experimental psychology, linguistics and communications engineering started to learn and integrate into their professional lives the kinds of mathematical skills that their elders had just barely become familiar with under the name of cybernetics.

Wiener has left an enduring monument, for in a sense many a young, tough-minded scientist and engineer from these fields is entitled to call himself today a "cyberneticien malgré lui."

MASSACHUSETTS INSTITUTE OF TECHNOLOGY
 (RESEARCH LABORATORY OF ELECTRONICS)

NORBERT WIENER AND POTENTIAL THEORY

BY M. BRELOT

1. Wiener did not work in potential theory for very long (only about two years around 1924), but that was enough to bring about in this field (the so-called classical potential theory), as in many others, some fundamental contributions: a definitive form of the generalized solution of the Dirichlet problem for a continuous given boundary function, the notion of capacity for general compact sets and the famous Wiener criterion of regularity.

2. Already in 1923, he wrote with Phillips a paper [28]* on *Nets and the Dirichlet problem* where this problem was solved for a polycubic domain then for domains with smooth boundaries by a limit process from a problem for functions on a discrete net and a mean condition. This idea of using linear equations for a preliminary problem relative to finite differences, which is now a basic tool with computers for partial differential equations was not common forty years ago; I know only the previous example of Le Roux on harmonic functions in R^2 (J. Math. Pures Appl., 1914).

3. In pure potential theory, the first fundamental paper of Wiener [24] *Certain notions in potential theory* in January 1924 gave and studied a precise definition of a generalized solution and the first definition of capacity for an arbitrary compact set.

For a long time, it was known that the classical Dirichlet problem does not always have a solution (case of an isolated boundary point of Zaremba, spine of Lebesgue) and there appeared more or less clearly the need to define a suitable generalized solution which always exists that would be later studied at the boundary; such a harmonic function corresponding to the given boundary function was in evidence in various methods, where further restrictions on the boundary allowed to show that it was actually a solution (Poincaré, Zaremba, Lebesgue, Bouligand, Kellogg . . .). But in a clearer and more striking way than the others, Wiener introduced for a bounded open set (and for a similar "exterior" problem) a precise generalized solution, *that he studied further without restrictions*: it was the limit of the classical solution for an increasing sequence of regular open sets $\Omega_n \subset \Omega$ ($\cup\Omega_n = \Omega$) ("regular" means that there is always a solution for the classical Dirichlet problem) and a boundary function, given as

---

* The bold-faced numbers in brackets refer to the numbered references in **the** Bibliography of Norbert Wiener.

the restriction on $\partial\Omega_n$ of any finite continuous continuation of the given function f on $\partial\Omega$. This limit depends only on Ω and f, and is equal to the classical solution when the latter exists.

Later an easy proof appeared by using the new tool of subharmonic function and an approximation of f by means of a difference $\phi_1-\phi_2$ of such functions in a larger $\Omega_0\supset\bar{\Omega}$, and considering the solutions in Ω_n for ϕ_1 and ϕ_2, which are increasing.

Wiener gave a rather complicated proof by considering an exterior problem relative to two compact sets and the respective boundary values 1 and 0 (at the point at infinity, the solution must tend to 0 in R^n, $n\geq3$, or must be bounded in R^2). Then he treated the more general case of n disjoint compact sets with constant values on them and that gave an approach to the general case where the boundary was replaced by n disjoint compact sets and a constant value on each one.

This difficult argument led him to consider more deeply the exterior problem with value 1 on the boundary of a given compact set K in R^n ($n\geq3$). Without the later representation of subharmonic functions by F. Riesz, the generalized solution was proved to be the newtonian potential of a measure ≥0 supported by K; the total mass was called the *capacity* of K, the basic tool of potential theory that Wiener adapted in R^2 and which has been so much used, studied and generalized. Moreover Wiener gave a sufficient condition of "regularity," broader than any previous ones. (This notion for a boundary point x means that the solution tends to the given value at x_0; the regularity of every boundary point means that the open set is regular.)

4. Three months later, in [35], [36], he studied more deeply this notion of regularity, already characterized by Lebesgue and his "barrier." Wiener considered first in R^n ($n\geq3$) the set of $C\Omega$ where the distance to x_0 lies between λ^p and λ^{p+1} ($0<\lambda<1$). If γ_p is its capacity, the regularity is equivalent to the divergence of the series γ_p/λ^p. This deep property (with an adaptation in R^2) was difficult to prove, but became easier later with the help of advanced potential theory. This famous criterion was systematically used and later extended to the more general notion of thinness; it appeared to have a geometric character according to the geometric interpretation of capacity by Fekete-Pólya-Szegö. (However in modern axiomatics, the barrier, which has not this character, remains valid but not the Wiener criterion.)

5. A third fundamental paper [39] was published in January 1925 where Wiener compared his generalized solution with the method of

Perron (published in Math. Z., vol. 18, where there is a quite similar and unknown paper of Remak). The method of Perron gave the classical solution (actually every time it exists) and Wiener proved that the function introduced by Perron, when considered before some complementary conditions make it the classical solution, was the generalized solution. The latter appears as the upper envelope of continuous subharmonic functions whose lim sup at the boundary minimize the given continuous data (and is a similar lower envelope). This remained valid with the further introduction of general subharmonic functions. This interpretation became the good definition and the first one a property of the generalized solution—with easy extension in the modern axiomatic theories.

The classical problem became therefore a particular case, with smaller interest.

6. Wiener could not leave aside the Dirichlet problem for a discontinuous boundary function. He was inspired by the Poisson-integral and considered the continuation of the functional defined at any $x \in \Omega$ by the solution for a given f on the boundary, when f becomes discontinuous, i.e., the integral of f with respect to the harmonic measure. (See [39] and previously [24].) He thought of course of the identity of the Perron envelopes for large classes of f. He surely also would have developed this point except for a simple counter-example that the present writer discovered to be wrong 15 years later (then the identity of the suitably defined envelopes was proved to be equivalent to the summability of f with respect to the harmonic measure).

In the papers I have just summarized, Wiener by mastering a difficult technique and introducing important tools initiated a new period for the Dirichlet problem and potential theory. We may be thankful for such an impulse that we may still appreciate 40 years later.

UNIVERSITY OF PARIS,
    PARIS, FRANCE

NORBERT WIENER ET L'ANALYSE DE FOURIER

J.-P. KAHANE

Il ne peut s'agir ici d'une revue complète de tout ce que Wiener a apporté à l'analyse de Fourier. Je voudrais seulement mettre l'accent sur son rôle de pionnier. Pour cela, quelques exemples suffiront.

1. La théorie générale des algèbres de Banach, élaborée par Gelfand, est souvent considérée aujourd'hui comme un fondement de l'analyse harmonique. Elle permet en tous cas d'aborder plus facilement, et de mieux comprendre, certaines questions comme les théorèmes taubériens de Wiener, dont parle ailleurs S. Mandelbrojt. Mais en vérité il n'est pas exagéré de dire que les algèbres de Banach sont issues du premier chapitre de *Tauberian theorems* [74].[1] L'étude qu'y fait Wiener des sommes de séries de Fourier absolument convergentes, et des transformées de Fourier de fonctions sommables sur la droite, a fourni un modèle aux travaux ultérieurs de Paul Lévy, de Beurling, et finalement de Gelfand. En même temps, les problèmes laissés en suspens ont stimulé des recherches pendant trente ans; rappelons seulement le célèbre "problème de Wiener," implicitement posé par l'énoncé du Théorème III de *Tauberian theorems*: soit f_1 et f_2 deux fonctions sommables sur la droite; est-il vrai que f_2 soit limite dans L^1 de combinaisons linéaires de translatées de f_1 dès que sa transformée de Fourier $\hat{f}_2$ s'annule partout où $\hat{f}_1$ s'annule? En termes plus abstraits: est-il vrai que tout idéal fermé de l'algèbre de convolution $L^1(R)$ soit l'intersection des idéaux maximaux qui le contiennent? On sait que la réponse, négative, n'a été apporté qu'en 1959, par Malliavin (la réponse pour $L^1(R^p)$, $p \geq 3$, ayant été découverte antérieurement par L. Schwartz).

2. Autre modèle d'algèbre de Banach: l'algèbre $\hat{M}$ des transformées de Fourier-Stieltjes des mesures bornées sur la droite. Les idéaux maximaux n'y sont pas si simples que dans les exemples précédents; c'est ce que montre l'exemple de Wiener et Pitt [105]: une $\hat{\mu} \in \hat{M}$, telle que $1/\hat{\mu}$ existe et soit bornée, mais n'appartienne pas à $\hat{M}$. Ç'a été là encore une belle source de travaux—le plus important restant le mémoire de Šreider de 1950.

3. Un sujet fascinant est la relation entre les propriétés d'une mesure bornée μ et le comportement à l'infini de sa transformée de

---

[1] Les nombres entre crochets renvoient à la bibliographie générale des oeuvres de Norbert Wiener.

Fourier $\hat{\mu}$. Un fait fondamental découvert par Wiener, et d'ailleurs facile, est que la mesure μ est continue (c'est-à-dire que sa primitive est continue) si et seulement si $|\hat{\mu}|^2$, ou aussi bien $|\hat{\mu}|$, a une valeur moyenne nulle sur la droite [34], [73, §5]. Avec Wintner [106], Wiener a construit d'autre part le premier exemple d'une mesure μ purement singulière telle que

$$\hat{\mu}(u) = O(|u|^{-1/2+\epsilon}) \qquad (u \to \pm \infty, \quad \epsilon > 0 \text{ donné});$$

la question était alors pendante depuis longtemps, et visiblement les auteurs ont eu une grande joie à la résoudre; ils ont même cru pouvoir construire la mesure μ de façon que son support soit un ensemble de mesure lebesguienne nulle, mais cela n'a été réalisé qu'ultérieurement, par Salem. Le problème est de savoir pour quelles fonctions $\phi(|u|)$ il existe une mesure μ purement singulière (ou mieux, une mesure $\mu \neq 0$ portée par un ensemble de mesure lebesguienne nulle), telle que $\hat{\mu}(u) = O(\phi(|u|))$ $(u \to \pm \infty)$. Il n'a été, aussi complètement que possible, résolu qu'en 1957, par Ivašev-Musatov: si ϕ décroit assez régulièrement, la condition nécessaire et suffisante pour qu'il en soit ainsi est $\int_0^\infty |\phi|^2 = \infty$.

4. Voici un autre aspect du sujet: quel est le comportement à l'infini de la transformée de Fourier $\hat{f}$ d'une fonction f s'annulant en dehors d'un intervalle? Wiener développe cette question pour donner une nouvelle démonstration du théorème de Denjoy-Carleman sur les fonctions quasi-analytiques: la classe des fonctions f telles que

$$\sup_{x,n} |f^{(n)}(x) A_n^{-1}| < \infty$$

est quasi-analytique si et seulement si

$$\int_0^\infty \log\left(\sum_{n=0}^\infty \frac{x^{2n}}{A_n^2}\right) \frac{dx}{1+x^2} = \infty \qquad ([92], \text{ Chapitre I}).$$

En liaison justement avec l'introduction de nouvelles classes quasi-analytiques et non quasi-analytiques (celles-ci étant destinées, par dualité, à fournir de nouvelles classes de distributions), le problème a été, au cours des dix dernières années, repris et étudié par Gelfand et Šilov et par Mandelbrojt sous la forme suivante: pour quels couples de fonctions ϕ et ψ, assez régulières, existe-t-il une $f \neq 0$ telle que $|f| \leq \phi$ et $|\hat{f}| \leq \psi$? Encore qu'il ait beaucoup progressé récemment (les dernières contributions étant dues à Katznelson et Mandelbrojt), le sujet est loin d'être épuisé.

5. Cela nous amène au cercle des idées développées dans le livre de Paley et Wiener: *Fourier transforms in the complex domain* [92]. D'abord le fameux théorème qui caractérise les transformées de Fourier des fonctions de carré sommable nulles hors d'un intervalle $[-a, a]$ comme fonctions de carré sommable qui sont restrictions à la droite réelle de fonctions entières de type exponentiel $\tau \leqq a$. Ce théorème est un outil très important pour de nombreuses questions d'analyse harmonique; outre celles, très nombreuses, qui figurent dans le livre de Paley et Wiener,[2] signalons seulement la théorie des fonctions moyennes périodiques, sous la forme que lui a donné L. Schwartz en 1947. Il a été étendu à plusieurs dimensions et traduit en termes de distributions c'est sous cette forme, par exemple, qu'il intervient dans la démonstration du "théorème des supports" de Lions: étant donné deux distributions S et T à supports compacts, le plus petit convexe portant la convolution $S * T$ (qu'on appelle son support convexe) est la somme des supports convexes de S et de T. Il a surtout attiré l'attention sur l'algèbre des fonctions entières de type exponentiel bornées sur la droite réelle, dont il convient de dire un mot maintenant.

6. Désignons par E l'algèbre des fonctions entières de type exponentiel, par E_b le sous-algèbre de E formée des fonctions bornées sur la droite réelle, et notons E^p et E_b^p les sous-algèbres des précédentes formées de fonctions paires. On voit facilement que E^p est l'ensemble des

$$
(1) \qquad f(z) = \prod_{1}^{\infty} \left(1 - \frac{z^2}{\lambda_n^2} \right), \qquad \lambda_n \text{ complexes,}
$$

$$
\lambda_n = O(n) \qquad (n \to \infty).
$$

Ainsi le quotient de deux fonctions $\in E^p$, si c'est une fonction entière,

---

[2] Qu'on se réfère pour cela à l'introduction, où Wiener célèbre à juste titre la puissance de la méthode:

"this work covered a great variety of topics, but was unified by the central idea of the application of the Fourier transform in the complex domain. . . . Nobody seems to have realized anything like the scope of the method. With its aid, we were able to attack such diverse analytic questions as those of quasi-analytic functions, of Mercer's theorem of summability, of Milne's integral equation of radiative equilibrium, of the theorems of Müntz and Szasz concerning the closure of sets of powers of an argument, of Titchmarsh's theory of entire functions of semi-exponential type with real negative zeros, of trigonometric interpolation and developments in polynomials of the form $\sum_{1}^{N} A_n \exp(i\lambda_n x)$, of lacunary series, of generalized harmonic analysis in the complex domain, of the zeros of random functions, and many others."

appartient encore à E^p; et la même chose vaut pour E, d'après un théorème de Lindelöf. Mais le quotient de deux fonctions $\in E_b^p$ peut être une fonction entière sans appartenir à E_b^p (par exemple, $z = (\sin^2 z/z)(\sin^2 z/z^2)^{-1}$), et la caractérisation des suites $\{\lambda_n\}$ telles que (1) définisse une fonction $\in E_b^p$ est hors de portée. Faute d'étudier E_b^p, Paley et Wiener introduisent une algèbre intermédiaire entre E_b^p et E^p, à savoir l'ensemble des fonctions (1) à zéros réels ($0 < \lambda_1 < \lambda_2 < \cdots$) satisfaisant l'une des propositions équivalentes qui suivent:

$$(2) \qquad \log f(iy) \sim \pi A(y) \qquad \text{quand } y \to \pm\infty,$$

$$(3) \qquad \int_{-\infty}^{\infty} \log |f(x)| \, \frac{dx}{x^2} = -\pi^2 A,$$

$$(4) \qquad \lim_{n \to \infty} \frac{n}{\lambda_n} = A.$$

La démonstration de l'équivalent (2)⇔(3)⇔(4) occupe une place centrale dans le livre de Paley et Wiener (Chapitre V). Un peu plus tard, Levinson devait serrer de plus près l'étude de E_b, mais les progrès décisifs en la matière ne viennent que des travaux récents de Beurling et Malliavin, dont voici un résultat marquant: si $f \in E_b$, $g \in E_b$, et $h = f/g \in E$, il existe une $k \in E_b$, de type exponentiel arbitrairement petit, telle que $hk \in E_b$.

7. La question qui précède est en étroite relation avec la suivante: on donne une suite d'exponentielles $\{\exp(i\lambda_n t)\}$; il existe un $l \geqq 0$ tel que pour tout intervalle I de longueur $< l$, ces exponentielles engendrent $L^2(I)$, et qu'il n'en soit plus ainsi dès que la longueur de I dépasse l. Comment calculer l quand on donne $\{\lambda_n\}$? Par dualité et transformation de Fourier, le problème revient à calculer la borne inférieure des types des fonctions $\in E_b$ qui s'annulent sur $\{\lambda_n\}$ (et $\not\equiv 0$).

Le cas $\lambda_n = i\mu_n$, $\mu_n > 0$, étudié dans le Chapitre II du livre, avait déjà fait l'objet des travaux de Müntz et de Szasz: alors $l = 0$ ou $l = \infty$ suivant que $\sum 1/\mu_n$ converge ou diverge. Le plus intéressant est l'étude du sous-espace de $L^2(I)$ engendré par les exponentielles $\{\exp(-\mu_n t)\}$ lorsque $l = 0$; le fait, fort intéressant, qu'il soit formé de fonctions analytiques a été découvert indépendamment aux alentours de 1940 par L. Schwartz, Clarkson et Erdös, et A. F. Leontiev.

Le cas où les λ_n sont réels occupe tout le Chapitre VI, mais le calcul explicite de l n'est fait que dans des cas particuliers. Ce n'est que dans le travail cité de Beurling et Malliavin que $l/2\pi$ apparait comme une "densité" effectivement calculable ("densité effective extérieure").

8. Dans quel cas $\{\exp(i\lambda_n t)\}$ $n = \cdots -1,\ 0,\ 1,\ \cdots$ est-elle une base dans $L^2(-\pi,\ \pi)$? Pour donner au moins des conditions suffisantes, et comparer les développements suivant les bases $\{\exp(int)\}$ et $\{\exp(i\lambda_n t)\}$, Paley et Wiener introduisent (Chapitre VII) la notion de bases $\{f_n\}$ et $\{g_n\}$ d'un espace de Hilbert, voisines dans le sens que

$$\sup \left\| \sum a_n(f_n - g_n) \right\| < 1$$

la borne supérieure étant prise pour toutes les suites complexes a_n telles que

$$\sum |a_n|^2 \leqq 1.$$

C'est là le point de départ de nombreux travaux, et en particulier de ce que Duffin et Schaeffer ont appelé la théorie des cadres (1952).

9. L'étude des développements $\sum a_n \exp(i\lambda_n t)$ est évidemment en rapport avec la théorie des fonctions presque périodiques. Dans ce domaine, une nouvelle classe de fonctions a été introduite par Paley et Wiener (Chapitre VII) sous le nom de fonctions pseudopériodiques: f est pseudopériodique si elle appartient à L^2 sur chaque intervalle borné et si les combinaisons linéaires de f et de ses translatées ont des normes uniformément équivalentes dans $L^2(I)$ et $L^2(J)$ quand I et J sont des intervalles égaux assez grands; la borne inférieure des longueurs permises pour I et J est appelée la pseudopériode et notée $L(f)$. L'intérêt de cette notion est que, pour toute une foule de propriétés, le fait pour f d'y satisfaire sur un intervalle de longueur supérieure à $L(f)$ entraine que f y satisfait partout. Paley et Wiener montrent que f est pseudopériodique si et seulement si f est presque périodique au sens de Stepanoff avec un spectre régulier (c'est-à-dire $\inf |\lambda_m - \lambda_n| = \delta > 0$), et ils appliquent la méthode à des séries trigonométriques lacunaires (dans le sens $\lim_{n \to \pm\infty} (\lambda_{n+1} - \lambda_n) = \infty$). Un peu plus tard et indépendamment, Ingham a donné une inégalité sur les polynômes trigonométriques qui améliore l'estimation de la pseudopériode donnée par Paley et Wiener. Le calcul exact de la pseudopériode date d'une dizaine d'années, et fait intervenir la "densité uniforme extérieure" de la suite $\{\lambda_n\}$. Jusqu'il y a peu de temps, la notion de pseudopériodicité n'avait guère attiré l'attention; il y a pourtant là une méthode intéressante et une bonne réserve de problèmes.

10. On connait mieux l'importante théorie que Wiener a développée sous le nom d'analyse harmonique généralisée; il s'agit essentiellement des fonctions mesurables f telles que

$$\phi(x) = \lim_{T \to \infty} \frac{1}{2T} \int_{-T}^{T} f(x + u)\, \overline{f(u)}\, du$$

existe pour tout x réel. Alors ϕ est transformée de Fourier d'une mesure positive, la mesure spectrale de f. Parmi elles figurent les fonctions presque périodiques de Bohr, mais aussi bien d'autres fonctions, en particulier les réalisations de nombreux processus stochastiques. Les idées développées dans le monumental article *General harmonic analysis* [73] ont, de ce fait, jeté un pont entre la théorie des processus stochastiques (mouvement brownien en particulier) et l'analyse de Fourier proprement dite. La portée des idées de Wiener en théorie des probabilités mérite une étude spéciale. Parmi celles, dans cet ordre de préoccupations, qui ont enrichi l'analyse harmonique, je me bornerai à un exemple [73, §12]. Pour démontrer l'existence d'une fonction f dont la mesure spectrale soit continue, Wiener procède ainsi: il choisit au hasard avec la probabilité naturelle, indépendamment les uns des autres, des $\epsilon_n = \pm 1$, et pose $f(x) = \epsilon_n$ sur $[n, n+1[$; alors, presque sûrement, la mesure spectrale de f est continue. Certes on peut construire des exemples explicites (Wiener en donne d'ailleurs un), mais la méthode introduite est générale et puissante, et elle a ensuite été appliquée dans des circonstances variées: au lieu de construire péniblement un contre-exemple, on construit un objet aléatoire dont presque toutes les réalisations constituent des contre-exemples.

Il faudrait encore signaler l'importance des idées de Wiener dans la théorie des équations intégrales (en particulier, celles du type de Wiener-Hopf); contentons nous de renvoyer à l'article de M. G. Krein paru dans les Uspehi en 1958.

Répétons qu'il ne s'agit là que de quelques exemples, destinés à dégager l'actualité de l'oeuvre de Wiener. La lecture des *Selected papers* [214] et des autres mémoires originaux, celle des livres de Wiener et en particulier du livre de Paley et Wiener, peuvent encore jouer un rôle de stimulant précieux en analyse de Fourier.

UNIVERSITY OF PARIS,
PARIS, FRANCE

LES TAUBÉRIENS GÉNÉRAUX DE NORBERT WIENER

S. MANDELBROJT

Il est peu probable que A. Tauber, qui donna le premier théorème réciproque (conditionnel, bien entendu) du théorème d'Abel, à savoir que $\lim_{x \to 1} \sum A_n x^n = S$, avec $A_n = o(1/n)$, implique $\sum A_n = S$, ait jamais pu songer que son nom serait donné à un ensemble aussi vaste de théorèmes de l'Analyse. Il est vrai que, quels que soient l'intérêt et l'importance des "taubériens" démontrés entre 1897—date de la publication du théorème de Tauber—et la parution des taubériens généreaux de Wiener, ce sont ces derniers qui ont élevé au rang d'une théorie structurée et harmonieuse l'ensemble des recherches consacrées à ce subjet.

Il semble que ce soit le théorème de Littlewood permettant de substituer le "O" au "o" de Tauber qui constitue la première étape importante "pre-wienerienne" dans cet ordre d'idées. Landau a pu remplacer la condition de Littlewood par la condition $nA_n > -K$, et celle-ci, jointe à la sommabilité au sens de Lambert de la suite A_n vers A, permet, d'après Hardy et Littlewood, d'affirmer que $\sum A_n = A$.

En partant de ce dernier taubérien, des auteurs ont pu déduire le théorème classique d'Hadamard et de de la Vallée-Poussin concernant la distribution des nombres premiers: $\pi(x) \sim x/\log x$.

Sans insister ici sur d'autres résultats très profonds de Landau, de Hardy-Littlewood et sur un théorème d'Ikehara, qui est une des plus belles applications des taubériens généraux, et dont nous parlerons plus loin, on aperçoit après cette simple énumération la richesse et les ramifications de l'ensemble des faits ainsi établis. Mais on s'aperçoit aussi de la dispersion de ces résultats, et des idées qui les ont guidés. On pouvait admirer le talent et la technique des chercheurs, mais il était difficile d'apercevoir une construction relevant d'une structure indiquant le fond de l'ensemble de ces recherches.

D'avoir trouvé un principe général concernant les transformées de Fourier, et notamment les convolutions—principe qui englobe les résultats cités et tant d'autres, anciens et nouveaux, tous importants —est le grand mérite de Wiener. Une des plus beaux chapitres de l'oeuvre de Norbert Wiener, de l'analyse harmonique, et de l'analyse, dite "classique," tout court, est ainsi créé.

Le taubérien général de Wiener est constitué en réalité de deux formes de théorèmes dont voici les deux énoncés essentiels [74], [81]:*

---

* Les nombres entre crochets renvoient à la bibliographie générale des oeuvres de Norbert Wiener.

THÉORÈME 1. *Soit $K_1 \in L$, et supposons que $\hat{K}_1(u) \neq 0$ pour tout u réel ($\hat{K}_1$ désigne la transformée de Fourier de K_1). Soit h une fonction bornée sur la droite. Si*

$$\lim_{x \to \infty} \int K_1(x - y)h(y)\, dy = A \int K_1(x)\, dx,$$

on a aussi

$$\lim_{x \to \infty} \int K(x - y)h(y)\, dy = A \int K(x)\, dx$$

pour tout K appartenant à L.

THÉORÈME 2. *Soient $K_1 \in L$, $K \in L$, les deux fonctions étant continues et supposons que*

$$\sum_{n=-\infty}^{\infty} \operatorname*{Sup}_{n \leq x < n+1} |K_1(x)| < \infty, \qquad \sum_{n=-\infty}^{\infty} \operatorname*{Sup}_{n \leq x < n+1} |K(x)| < \infty,$$

avec $\hat{K}_1(u) \neq 0$ pour tout u réel. Soit h une fonction localement à variation bornée, telle que pour tout x

$$\int_x^{x+1} |dh(y)| < M.$$

Si

$$\lim_{x \to \infty} \int K_1(x - y)\, dh(y) = A \int K_1(x)\, dx,$$

on a aussi

$$\lim_{x \to \infty} \int K(x - y)\, dh(y) = A \int K(x)\, dx.$$

La démonstration originale du Théorème 1 est basée sur le théorème suivant concernant le sous-espace engendré par les translatés de K_1, théorème qui joue, lui-même, un rôle extrêmement important dans l'analyse harmonique.

THÉORÈME 3. *Si $K_1 \in L$, $K \in L$, $\hat{K}_1(u) \neq 0$ pour tout u réel, à tout $\epsilon > 0$ correspond une suite A_n et une suite de nombres réels ξ_n, les deux suites finies, telles que*

$$\int \big|\, K(x) - \sum A_n K_1(x + \xi_n)\,\big|\, dx < \epsilon.$$

D'ailleurs, la condition $\hat{K}_1(u) \neq 0$ est aussi nécessaire pour qu'une telle approximation soit possible pour tout K de L.

Le Théorème 3 est un cas particulier d'un théorème beaucoup plus vaste. Ainsi [81]:

THÉORÈME 4. *Si $K \in L$ et si $\sum$ est une sous-classe de L, dont les transformées de Fourier n'ont pas de zéro commun (réel), K peut être approximé par des combinaisons linéaires finies des fonctions de $\sum$.*

Désignons par A la classe des fonctions qui sont les transformées de Fourier des fonctions de L (plusieurs auteurs dénotent cette classe, très justement, par W). Une des étapes essentielles dans la démonstration des théorèmes cités (aussi bien dans la démonstration originale de Wiener du Théorème 3 que dans celles des auteurs qui ont fourni d'autres démonstrations pour théorème 1) consiste à démontrer que la division par une fonction de la classe A fournit, dans un intervalle où cette transformation ne s'annule pas, la restriction d'une fonction de la même classe. Voici exactement ce théorème (dont la version "séries trigonométriques" remonte à Denjoy):

THÉORÈME 5. *Si $k \in A$ avec $k(u) \neq 0$ sur $[a, b]$, et si $f \in A$ avec $f(u) = 0$ pour $u \notin [a, b]$, on a aussi $f/k \in A$.*

C'est à partir de ce théorème qu'a pu prendre naissance le théorème appelé "théorème de Lévy-Wiener" indiquant l'analyticité de F comme condition suffisante pour que cette fonction "opère" dans A —résultat qui a provoqué des recherches très importantes durant la dernière décennie.

Un des passages du taubérien général de Wiener au théorème sur la distribution des nombres premiers consiste nécessairement (que ce passage soit fait par Wiener ou par ceux qui ont utilisé son taubérien) en la constatation que la transformée de Fourier d'une certaine fonction K_1 $(\in L)$ ne s'annule pas; or, il est intéressant de constater que dans ce passage la transformée de Fourier en question se trouve toujours être, à des facteurs simples près (pour que le produit soit la transformée de Fourier d'une fonction de L), la fonction $\zeta(1 + iu)$.

Est-ce la recherche d'une nouvelle démonstration du théorème d'Hadamard-de la Vallée-Poussin, démonstration dans laquelle le non-évanouissement de $\zeta(s)$ sur $\Re s = 1$ se traduit par le non-évanouissement de la transformée de Fourier d'une fonction, facteur d'une convolution, qui a conduit l'auteur à son idée extrêmement générale?

C'est en cherchant à se débarrasser de quelques hypothèses intervenant dans un théorème de Landau, et dans un théorème bien plus général de Hardy et Littlewood, hypothèses manifestement peu appropriées au sujet, que S. Ikehara a obtenu son théorème bien connu. Et c'est parce que ce théorème, particulièrement intéressant, est obtenu par l'auteur comme une des premières applications du taubérien général de Wiener (sous Théorème 2) que nous croyons utile de l'énoncer ici:

$\alpha(y)$ étant une fonction non décroissante pour $y > 1$, si

$$A(w) = \int_{1+0}^{\infty} y^{-w} d\alpha(y)$$

converge absolument pour $\Re w > 1$, et si la fonction $A(w) - (w-1)^{-1}$ est continue pour $\Re w \geqq 1$, on a $\alpha(y) \sim y (y \to \infty)$.

Ce résultat a conduit Wiener à un théorème qui, une fois de plus, ressemble fortement au passage de $\zeta(1+iu) \neq 0$ au théorème $\pi(x) \sim x/\log x$ [81].

Avec les notations précédentes, supposons que la fonction $e^{A(w)}(w-1)^A$ $(0 < A < 2\frac{1}{2})$ puisse être prolongée analytiquement sur $\Re w = 1$, sans s'annuler sur cette droite. On a alors

$$A = \lim_{N \to \infty} \frac{1}{N} \int_{1+0}^{N} \log x \, d\alpha(x).$$

Comme Wiener l'a démontré, les conclusions des Théorèmes 1 et 3 restent valables si $\hat{K}_1(u)$ ne s'annule que dans des intervalles fermés intérieurs aux intervalles où $\hat{K}(u) = 0$ [74].

C'est à partir de ce point que de nombreux problèmes ont été posés, problèmes qui conduisent à la synthèse harmonique moderne.

Terminons en remarquant qu'un théorème du type 3 a été également démontré par Wiener dans L_2, théorème où la condition $\hat{K}_1(u) \neq 0$ est remplacée, bien entendu, par la condition que $\hat{K}_1(u) = 0$ sur un ensemble de mesure nulle.

COLLÈGE DE FRANCE,
    PARIS, FRANCE

WIENER AND INTEGRATION IN FUNCTION SPACES

BY M. KAC

1. Evolution of Mathematics is, by and large, a continuous process and its growth and progress seldom deviate greatly from the natural historical lines. It is because of this that we tend, in retrospect, to admire most those developments which though born well outside it have grown to join and to enrich the mainstream of our science.

It was the great fortune and the great achievement of Norbert Wiener to initiate such a development when, in the early twenties, he introduced a measure, now justly bearing his name, in the space of continuous functions.

2. Let us first review briefly some of the background.

At about 1905, almost simultaneously, and quite independently of each other (in fact, using wholly different approaches) A. Einstein and M. Smoluchowski provided a theory of the peculiar erratic motion of small particles suspended in liquids first described in 1828 by the English botanist Brown.

The theory can be summarized as follows:

(a) For simplicity one confines one's attention to the displacement of the Brownian particle in some chosen direction and one can thus speak of the one-dimensional Brownian motion.

(b) The motion is Markoffian and homogeneous in time; i.e. the probability of finding the particle at times $t_1, t_2, \cdots, t_n$ $(0 < t_1 < t_2 < \cdots < t_n)$ in the intervals $(\alpha_1, \beta_1), \cdots, (\alpha_n, \beta_n)$ is given by the formula

$$(2.1) \quad \int_{\alpha_1}^{\beta_1} \int_{\alpha_2}^{\beta_2} \cdots \int_{\alpha_n}^{\beta_n} P(x_0 \mid x_1; t_1) P(x_1 \mid x_2; t_2 - t_1) \cdots$$

$$\cdots P(x_{n-1} \mid x_n; t_n - t_{n-1}) dx_1 \cdots dx_n$$

where $P(x \mid y; t)$ is the probability density of finding the particle at y at time t if it started at x at $t = 0$.

(c) For $\Delta t \to 0$ one has asymptotically

$$\langle \Delta x \rangle = \int_{-\infty}^{\infty} (x - x_0) P(x_0 \mid x; \Delta t) \, dx \sim F(x_0) \Delta t,$$

$$\langle (\Delta x)^2 \rangle = \int_{-\infty}^{\infty} (x - x_0)^2 P(x_0 \mid x; \Delta t) \, dx \sim 2D\Delta t,$$

$$\langle |\Delta x|^k \rangle = o(\Delta t), \qquad k \geq 3.$$

Here $F(x)$ is the *outside force* (e.g. gravity) acting on the particle when it is at x, and D (the diffusion constant) is given by the formula

$$D = \frac{2kT}{f}$$

where T is the absolute temperature, f the friction coefficient (in liquids e.g. f is given by Stokes' formula $f = 6\pi a\eta$ with a the radius of the spherical particle and η the viscosity coefficient) and k the Boltzmann constant.

From (b) it follows at once that $P(x|y; t)$ must satisfy the Smoluchowski equation (often called the Chapman-Kolmogoroff equation),

$$(2.2) \qquad P(x|y; t + \tau) = \int_{-\infty}^{\infty} P(x|\xi; t) P(\xi|y; \tau) \, d\xi, \qquad \tau > 0,$$

and from this, using (c) one can derive (under appropriate smoothness conditions on P) the diffusion equation

$$\frac{\partial P}{\partial t} = D \frac{\partial^2 P}{\partial x^2} + \frac{\partial}{\partial x} (F(x)P).$$

We have, in addition, the initial condition

$$P(x|y; t) \to \delta(y - x), \qquad \text{as } t \to 0,$$

and the obvious restriction that

$$P(x|y; t) \geq 0.$$

If e.g. $F(x) = 0$, i.e. we are dealing with a free Brownian particle, the unique solution of the problem is

$$(2.3) \qquad P(x|y; t) = \frac{1}{2(\pi Dt)^{1/2}} \exp\left(-\frac{(y - x)^2}{4Dt}\right)$$

and, in particular, the mean-square displacement at time t is given by the formula

$$\langle \Delta^2(t) \rangle = \int_{-\infty}^{\infty} (y - x)^2 P(x|y; t) \, dy = 2Dt.$$

Thus if one observes a large number of free Brownian particles during the same time interval t and equates the empirically calculated mean-square derivation with the theoretically predicted value $2Dt$, one can get an estimate of D and hence determine (empirically and approximately) the Avogadro number.

The successful determination of the Avogadro number from Brownian motion experiments was one of the great triumphs of Physics in the early days of the century and it dealt the final blow to the opponents of atomistic theories.

By the time Smoluchowski died in 1917 the theory of Brownian motion was accepted, understood and, in many respects, finished.

3. According to his own account, (see [177]),[1] Wiener's concern with the problem of measure in function spaces began during his first term as an Instructor at Massachusetts Institute of Technology and it was Professor I. Barnett, of the University of Cincinnati, who drew his attention to the interest of "generalization of the concept of probability to cover probabilities where the various occurrences being studied were not represented by points or dots in a plane or in space but by something of the nature of path curves in space" [177, p. 35].

I find it enormously surprising that, in 1919, when probability theory was not even thought of as a branch of pure mathematics, two young men should have contemplated problems of such degree of sophistication!

Be as it may, Wiener became preoccupied with the subject of measure and integration in a space of curves and in the process he familiarized himself with the work of Gâteaux and especially with that of P. J. Daniell. At the same time he also read G. I. Taylor's pioneering paper on turbulence and he came to think of "the physical possibilities of a theory for averages over curves" [177, p. 37].

"The problem of turbulence," Wiener continues in his autobiography [177, pp. 37, 38, 39] "was too complicated for immediate attack,[2] but there was a related problem which I found to be just right for the theoretical considerations of the field I had chosen for myself. This was the problem of the Brownian motion, and it was to provide the subject of my first major mathematical work. . . . Here I had a situation in which particles describe not only curves but statistical assemblages of curves. It was an ideal proving ground for my ideas concerning the Lebesgue integral in a space of curves, and it had the abundantly physical texture of the work of Gibbs. It was to this field that I had decided to apply the work that I had already done along the lines of integration theory. . . ."

---

[1] The bold-faced numbers in brackets refer to the numbered references in the Bibliography of Norbert Wiener.

[2] It still is today!

"The Brownian motion was nothing new as an object of study by physicists. There were fundamental papers by Einstein and Smoluchowski that covered it, but whereas these papers concerned what was happening to any given particle at a specific time, or long-time statistics of many particles, they did not concern themselves with the mathematical properties of the curve followed by a single particle.

"Here the literature was very scant, but it did include a telling comment by the French physicist Perrin in his book *Les Atomes*, where he said in effect that the very irregular curves followed by particles in the Brownian motion, led one to think of the supposed continuous nondifferentiable curves of the mathematicians."

4. Wiener presented his basic ideas in a series of papers published in the period from 1920 to 1923. He included an account of them in his famous Acta paper on generalized harmonic analysis and he devoted a chapter of his 1934 book with Paley to presenting again the subject to the mathematical community.

The early papers were very difficult to read (as I know from personal experience, when as a student in Lwów I tried to read them with a depressing lack of success) and they fell on deaf ears. Only Paul Lévy in France, who had himself been thinking along similar lines, fully appreciated their significance.

The Acta paper and the book on the Fourier transform in the complex domain were so full of new and exciting results and ideas, which were so unquestionably in the mainstream of analysis that the esoteric "random functions" were somehow overlooked.

The original problem which Wiener posed himself was the following:

Can one introduce in the space of *continuous functions* $x(t)$ $(0 \leq t < \infty, x(0) = 0)$ a *completely additive* measure μ such that the measure of the set of paths which at times $t_1, t_2, \cdots, t_n$ $(0 \leq t_1 < t_2 < \cdots < t_n)$ pass through the "gates" $(\alpha_1, \beta_1), (\alpha_2, \beta_2), \cdots, (\alpha_n, \beta_n)$ is given by the Einstein-Smoluchowski formula (2.1) with

$$(4.1) \qquad P(x \mid y; t) = \frac{1}{(2\pi t)^{1/2}} \exp\left(-\frac{(y-x)^2}{2t}\right)?$$

In other words we require that μ be such that

$$\mu\{\alpha_1 < x(t_1) < \beta_1, \cdots, \alpha_n < x(t_n) < \beta_n\}$$

$$(4.2) \qquad = \int_{\alpha_1}^{\beta_1} \cdots \int_{\alpha_n}^{\beta_n} P(0 \mid x_1; t_1) P(x_1 \mid x_2; t_2 - t_1) \cdots$$

$$\cdots P(x_{n-1} \mid x_n; t_n - t_{n-1}) \, dx_1 \cdots dx_n$$

with P given by (4.1). (Note that we set $D=\frac{1}{2}$ since the physical nature of the diffusion constant is of no relevance to this phase of the theory.)

Wiener's solution of this problem is embodied in the following theorem:

If G_n, $n=0, 1, 2, \cdots$, is a sequence of independent, normally distributed random variables each having mean 0 and variance 1, then the series

$$(4.3) \qquad \frac{G_0}{\pi^{1/2}}t + \sum_{n=1}^{\infty} \sum_{k=2^{n-1}}^{2^n-1} G_k \left(\frac{2}{\pi}\right)^{1/2} \frac{\sin kt}{k}$$

converges *uniformly* in $0 \leq t < \pi$, with probability 1, and denoting its sum by $x(t)$ we have (4.2) if we interpret μ as *probability*, or equivalently, as the product measure in the product

$$R^{\infty} = R \times R \times R \times \cdots$$

with R the real line and the measure ν in R defined by the formula

$$\nu(E) = \frac{1}{(2\pi)^{1/2}} \int_E \exp\left(-\frac{x^2}{2}\right) dx.$$

Thus to every continuous function $x(t)$ on the interval $0 \leq t < \pi$ and such that $x(0)=0$ there corresponds a point $(G_0, G_1, G_2, \cdots)$ in R^{∞} and conversely, to *almost* every point in R^{∞} (in the sense of the product measure defined above) there corresponds a continuous function $x(t)$ ($x(0)=0$) on $(0, \pi)$. Clearly, the almost one-to-one mapping

$$x(t) \rightarrow (G_0, G_1, G_2, \cdots)$$

can be used to define a measure on the space of continuous functions and while the construction is limited to functions defined on the interval $(0, \pi)$ it can be easily modified to apply to functions defined on $(0, \infty)$.

Having constructed the measure Wiener then proved that *almost every $x(t)$ is nowhere differentiable* and, even more strongly, that for every $\epsilon > 0$ almost every $x(t)$ satisfies the Lipshitz condition with exponent $\frac{1}{2}-\epsilon$ but almost none with exponent $\frac{1}{2}+\epsilon$.

Thus has the remark of Perrin turned into a beautiful theorem!

Let us recall now that at about 1930 there appeared a series of papers by Paley and Zygmund in which, following the lines suggested by the earlier work of Steinhaus and Kolmogoroff and Khintchine, trigonometric series with "random" plus-minus signs

$$\sum \pm c_n \cos nt$$

were extensively studied. Wiener's results appeared to belong naturally to this circle of problems and indeed Paley and Zygmund joined forces with Wiener in [87] which, in a way, was a "merger" of the two lines of development.

Because of this merger the original motivation was all but lost and what Wiener did took on the appearance of an elegant but special result concerning Fourier series with random coefficients. This subject, by the way, has shown remarkable vitality and there is a significant revival of interest in it largely because of the work of the late Salem and that of Kahane, Katznelson, Malliavin and others.

5. In the meantime measure theoretic foundations of probability theory became firmly established and toward the end of the nineteen-thirties it became possible to introduce in a precise and rigorous way very general stochastic processes. This was done mainly by J. L. Doob and while his methods and approach were quite different from (and in some respects preferable to) those of Wiener the basic motivation was, in principle, quite close to Wiener's original one. There were however subtle differences and it is perhaps worthwhile to discuss them briefly especially since in the process we can bring out several other relevant points.

Consider for example the problem of finding the probability that a free Brownian particle starting from $x=0$ will remain to the left of $x=a$ $(a>0)$ for all times τ not exceeding t.

The problem would present no difficulty to a physicist who would, following Smoluchowski, argue as follows.

If a large number of identical particles are started at $x=0$ and watched en masse the macroscopic appearance will be that of classical diffusion; consequently the desired probability is simply the proportion of particles which up to time t did not pass the mark $x=a$. It is thus sufficient to place an *absorbing barrier* at $x=a$ and calculate the fraction of the diffusing matter to the left of the absorbing barrier at time t. This is a classical problem and the answer, as is well known, can be obtained as follows:

Let $Q_a(0\,|\,x;t)$ be the fundamental solution of the diffusion equation

$$(5.1) \qquad \frac{\partial Q}{\partial t} = \frac{1}{2}\frac{\partial^2 Q}{\partial x^2}$$

subject, in addition to the obvious initial condition

$$(5.2) \qquad \lim_{t\to 0} Q(0\,|\,x;t) = \delta(x),$$

also to the boundary condition

$$(5.3) \qquad\qquad Q(0 \,|\, a; t) = 0.$$

The desired fraction of the diffusing material is then

$$(5.4) \qquad\qquad q(a; t) = \int_{-\infty}^{a} Q_a(0 \,|\, x; t)\, dx.$$

Since (for $x \leq a$)

$$Q_a(0 \,|\, x; t) = \frac{1}{(2\pi t)^{1/2}} \exp\left(-\frac{x^2}{2t}\right) - \frac{1}{(2\pi t)^{1/2}} \exp\left(-\frac{(x-2a)^2}{2t}\right),$$

we obtain

$$(5.5) \qquad\qquad q(a; t) = \frac{2}{(2\pi t)^{1/2}} \int_{0}^{a} \exp\left(-\frac{x^2}{2t}\right) dt.$$

In Wiener's setting the problem is to find the measure of the set of those paths $x(\tau)$ $(x(0)=0)$ which satisfy the condition

$$(5.6) \qquad\qquad x(\tau) \leq a, \qquad 0 \leq \tau \leq t.$$

The explicit representation (4.3) is useless for the purposes of actually computing the measure but since one *knows* that the paths are *continuous*, the set defined by (5.6) is measurable and moreover

$$(5.7) \quad \mu\{x(\tau) \leq a; 0 \leq \tau \leq t\} = \lim_{n \to \infty} \mu\left\{x\left(\frac{kt}{n}\right) \leq a; k = 1, 2, \cdots, n\right\}.$$

Using (4.2) we get at once that

$$\mu\left\{x\left(\frac{kt}{n}\right) \leq a; k = 1, 2, \cdots, n\right\}$$

$$(5.8) \qquad = \int_{-\infty}^{a} \cdots \int_{-\infty}^{a} P\left(0 \,|\, x_1; \frac{t}{n}\right) P\left(x_1 \,|\, x_2; \frac{t}{n}\right) \cdots$$

$$\cdots P\left(x_{n-1} \,|\, x_n; \frac{t}{n}\right) dx_1 \cdots dx_n$$

with

$$P(x \,|\, y; t) = \frac{1}{(2\pi t)^{1/2}} \exp\left(-\frac{(y-x)^2}{2t}\right).$$

The *existence* of the limit in (5.7) is a simple consequence of *measurability* of the set (5.6) and there remains "only" the *analytic* problem of proving that the limit is indeed equal to $q(a; t)$.

In Doob's approach one encounters an interesting difficulty at an earlier stage. Doob begins with the space of *all* real valued functions $x(\tau)$ (subject to the trivial normalization $x(0)=0$) but he keeps (4.2) as the assignment of measures to sets of functions satisfying the conditions

$$(5.9) \quad \alpha_1 < x(t_1) < \beta_1, \ \alpha_2 < x(t_2) < \beta_2, \cdots, \alpha_n < x(t_n) < \beta_n.$$

The Borel field generated by sets (5.9) turns out to be too small to make many of the interesting sets measurable. In particular, the set (5.6) is not measurable and neither, for that matter, is the set C of all continuous functions. Fortunately the outer measure of C turns out to be 1 (while the inner measure is 0) and because of this it is *possible* to concentrate the measure on it while maintaining (4.2). In this way one comes back to the Wiener measure but it now emerges as only one of an infinitude of possible measures consistent with (4.2).

The relative delicacy of the situation tended to polarize the interest in stochastic processes along measure-theoretic rather than analytic lines. One might say that more attention was paid to *defining* measures like

$$(5.10) \qquad x(\tau) \leq a; \qquad 0 \leq \tau \leq t$$

than to ways and means of *calculating* them.

But why fuss over measure-theoretic points when all the time the answer could be obtained (as shown above) by a simple application of classical diffusion theory?

The answer lies somewhat deeper than one might think.

It can be verified by an elementary computation that (5.4) and (5.5) imply for $x \leq a$ the equation

$$(5.11) \quad Q_a(x;t) - P(0 \,|\, x;t) - \int_0^t \left(-\frac{\partial q(a;\tau)}{\partial \tau} \right) P(a \,|\, x;t-\tau) \, d\tau,$$

or equivalently

$$\int_I Q_a(x;t) \, dx + \int_0^t \left(-\frac{\partial q(a;\tau)}{\partial \tau} \right) \int_I P(a \,|\, x;t-\tau) \, dx d\tau$$

$$(5.12)$$

$$= \int_I P(0 \,|\, x;t) \, dx,$$

where I is an interval to the left of $x=a$.

Note now that

$$-\frac{\partial q(a;\tau)}{\partial \tau} \, d\tau$$

is the (differential) probability that the Brownian path $x(t)$ $(x(0)=0)$
will leave the half line $x \leq a$, *for the first time*, between τ and $\tau+d\tau$,
and consequently

$$- \frac{\partial q(a; \tau)}{\partial \tau} d\tau \int_I P(a \mid x; t - \tau) \, dx$$

is the probability that the path will end up in I at time t having
crossed $x=a$ for the first time between τ and $\tau+d\tau$. The second integral
on the right hand side of (5.12) is thus the probability of ending up
in I at time t having left $(-\infty, a)$ sometime before, and the first
integral is clearly the probability of ending up in I at time t without
having left that half line; together they add up to the probability of
ending up in I at time t and equation (5.12) follows.

But in arriving at this appealing interpretation I have identified
leaving $(-\infty, a)$ and *crossing* $x=a$; this identification is justified *only
if the paths are continuous* and we are again led to the question of continuity of paths!

To appreciate this more fully let us imagine for a moment that the
displacement of a particle was governed by the Cauchy density

$$(5.13) \qquad P(x \mid y; t) = \frac{t}{\pi} \frac{1}{t^2 + (x - y)^2}$$

instead of the Gaussian density (4.1).

This density also satisfies the Smoluchowski equation (2.2) and
hence it is consistent to use formula (4.1) to assign measures to sets
(5.9).

Probability $q(a; t)$ could still be defined by formulas (5.7) and (5.8)
with P given by (5.13) and $Q_a(x; t)$ can be defined by the formula

$$Q_a(x; t) = \lim_{n \to \infty} \int_{-\infty}^a \cdots \int_{-\infty}^a P\left(0 \mid x_1; \frac{t}{n}\right) P\left(x_2 \mid x_3; \frac{t}{n}\right) \cdots$$
$$(5.14)$$
$$\cdots P\left(x_{n-1} \mid x; \frac{t}{n}\right) dx_1 \cdots dx_{n-1}$$

again with P being the Cauchy density (5.13). The limits involved in
(5.7) and (5.14) can be shown to exist and we might think that the
analogy with the Gaussian case is so complete that (5.11) also holds.
This however is not so!

The reason is that no longer can the measure be concentrated on the
space of continuous functions and therefore "leaving $(-\infty, a)$" and
"crossing $x=a$" are not the same. In fact, because of the discontinu-

ous nature of the paths "crossing $x = a$" is not even properly defined.

That the the sample paths of the Cauchy process (and other so-called stable processes) cannot be continuous (i.e. assigning the measure (4.1) with P given by (5.13) to sets (5.9) makes it *impossible* to concentrate the measure on the space C of continuous functions[3]) was first noted by Paul Lévy who has also shown that the sample paths, though by necessity discontinuous, can be made right (or left) continuous.[4] This development helped to underscore the depth of Wiener's achievement although full awareness of capabilities and potentialities of measure and integration theories in function spaces was rather slow in coming.

6. During the war years there has been a substantial increase of interest in stochastic processes owing mainly to the need of analyzing noise phenomena in radar and related electronic systems.

Here methods based on trigonometric series with random coefficients are especially appealing and natural.

Wiener always had a strong interest in circuits and in other phases of electrical engineering and he continued almost to the end of his life to relate his early work on "random functions" to linear and non-linear problems in circuit theory (see e.g. [191]).

But the strongest impact of Wiener's ideas on mathematics came from different directions.

Starting in 1943, R. H. Cameron and W. T. Martin published a series of papers devoted to the calculation of Wiener integrals of a class of functionals.

Their work culminated in proving that if $p(t) \geq 0$, $0 \leq t \leq 1$, is continuous, then the Wiener integral of

$$(6.1) \qquad \exp\left(\lambda \int_0^1 p(t) x^2(t)\, dt\right)$$

is equal, for sufficiently small λ, to

$$(6.2) \qquad \prod_{k=1}^{\infty} \frac{1}{(1 - \lambda\lambda_k)^{1/2}}$$

where the λ_k's are the eigenvalues of the Sturm-Liouville problem

$$(6.3) \qquad \phi'' + \frac{1}{\lambda} p(t)\phi = 0, \qquad \phi(0) = \phi'(0) = 0.$$

---

[3] In fact, in Doob's formulation described above, the outer measure of C is 0.

[4] Doob has greatly extended and systematized this work.

To derive this result Cameron and Martin developed an interesting theory of linear "changes of variable" in Wiener space in which linear integral equations played an important part.

This was the first inkling that there are highly nontrivial connections between classical analysis and integration in function spaces.

Almost at the same time Kakutani published a series of notes on connections between potential theory and Wiener measure. Of his results perhaps the most important was one identifying sets of capacity zero with those which Brownian motion curves hit with probability zero. This was a beginning of an extensive and fruitful development which continues to this day.

Then in 1948 there appeared in print a part of the 1942 doctoral dissertation of R. P. Feynman which dealt with a highly suggestive reformulation of the nonrelativistic quantum mechanics in terms of a certain integral over the space of paths.

To review briefly what is at stake let us restrict ourselves to the case of a particle of mass m moving along the x-axis in a force field generated by the potential $V(x)$.

The basic problem is to find the probability that if at time 0 the system is in the state described by the wave function ϕ_0, it will be in the state described by the wave function ϕ_1 at time t. This probability is of the general form

$$(6.4) \qquad |\langle \phi_0 | K | \phi_1 \rangle|^2$$

where K is an appropriate operator. In our one-dimensional case,

$$(6.5) \qquad \langle \phi_0 | K | \phi_1 \rangle = \int_{-\infty}^{\infty} \int_{-\infty}^{\infty} \phi_0(x_0) K(x_0 | x_2; t) \phi_1(x) \, dx_0 dx$$

where the "propagator" $K(x_0 | x; t)$ is obtained by solving the Schroedinger equation

$$(6.6) \qquad -\frac{\hbar}{i} \frac{\partial K}{\partial t} = -\frac{\hbar^2}{2m} \frac{\partial^2 K}{\partial x^2} + V(x)K$$

subject to the initial condition

$$(6.7) \qquad K(x_0 | x; t) \to \delta(x - x_0).$$

What Feynman noted was that K can be formally written as a certain average over the set of all paths $x(\tau)$ such that

$$(6.8) \qquad x(0) = x_0, \qquad x(t) = x$$

of the quantity

$$(6.9) \qquad \exp\left(\frac{i}{\hbar} S[x]\right)$$

where

$$(6.10) \qquad S[x] = \int_0^t \left\{ \frac{m}{2}\left(\frac{dx}{d\tau}\right)^2 - V(x(\tau)) \right\} d\tau$$

is the classical action integral (the integrand is the Lagrangean of the system). Symbolically,

$$(6.11) \qquad K(x_0 \mid x; t) = \int \exp\left(\frac{i}{\hbar} S[x]\right) d\,(\text{path}).$$

Feynman proposed to define his "integral" (6.11) as the limit

$$(6.12) \quad \lim_{n\to\infty} \frac{\displaystyle\int_{-\infty}^{\infty}\cdots\int_{-\infty}^{\infty} \exp\left[\frac{i}{\hbar}\Delta_n \sum_{k=0}^{n-1}\left\{\left(\frac{m}{2}\left(\frac{x_{k+1}-x_k}{\Delta_n}\right)\right)^2 - V(x_k)\right\}\right] dx_1 dx_2 \cdots dx_{n-1}}{\displaystyle\int_{-\infty}^{\infty}\cdots\int_{-\infty}^{\infty} \exp\left[\frac{i}{\hbar}\Delta_n \sum_{k=0}^{n-1}\frac{m}{2}\left(\frac{x_{k+1}-x_k}{\Delta_n}\right)\right] dx_1 dx_2 \cdots dx_{n-1}}$$

where $\Delta_n = t/n$, and $x_n = x$. Leaving aside questions of convergence he then showed (formally) that if K is defined in this way it is indeed the correct propagator i.e. it satisfies (6.6) and (6.7).

In 1947 Feynman and I were colleagues at Cornell and I attended a lecture of his at the Physics Colloquium at which he presented his formulation of quantum mechanics. At the time I happened to be interested in calculating various Wiener integrals (some suggested by discussions with W. T. Martin to whom I owe a great deal for getting me involved with the subject of integration in function spaces) and as I sat listening to Feynman it occurred to me that if instead of (6.6) one were to take the real heat equation

$$(6.13) \qquad \frac{\partial Q}{\partial t} = \frac{1}{2}\frac{\partial^2 Q}{\partial x^2} - V(x)Q$$

(still subject to the initial condition $Q(x_0\mid x, t) \to \delta(x-x_0)$ as $t\to 0$), then one could write its solution (Green's function) as an appropriate Wiener integral.

64   M. KAC

In fact, if $V(x)$ is bounded from below and satisfies mild regularity
conditions and if $f(x)$ is integrable, then

$$\int_{-\infty}^{\infty} Q(x_0 \mid x; t) f(x) \, dx$$

(6.14)

$$= E\left\{ \exp\left(- \int_0^t V(x_0 + x(\tau)) \, d\tau \right) f(x_0 + x(t)) \right\}$$

where $E\{ \cdots \}$ denotes the Wiener integral of the functional inside
the braces. Q itself could be obtained either by an appropriate limit
process or, equivalently, by considering *conditional* Wiener integrals[5]
(expectations).

The formula then is

$$Q(x_0 \mid x; t) = \frac{\exp\left(- \dfrac{(x - x_0)^2}{2t} \right)}{(2\pi t)^{1/2}}$$

(6.15)

$$\cdot E\left\{ \exp\left(- \int_0^t V(x_0 + x(\tau)) \, d\tau \right) \mid x(t) = x - x_0 \right\}.$$

If one were to follow Feynman, one would try to define Q (in analogy with (6.12)) as the limit

(6.16)
$$\lim_{n \to \infty} \frac{\displaystyle\int_{-\infty}^{\infty} \cdots \int_{-\infty}^{\infty} \exp\left[-\Delta_n \sum_{k=0}^{n-1} \left\{ \frac{1}{2} \left(\frac{x_{k+1} - x_k}{\Delta_n} \right)^2 + V(x_k) \right\} \right] dx_1 \cdots dx_{n-1}}{\displaystyle\int_{-\infty}^{\infty} \cdots \int_{-\infty}^{\infty} \exp\left[-\Delta_n \sum_{k=0}^{n-1} \frac{1}{2} \left(\frac{x_{k+1} - x_k}{\Delta_n} \right)^2 \right] dx_1 \cdots dx_{n-1}},$$

but now, owing to the *rigorously established existence of the Wiener
measure*,[6] the existence of the limit is an immediate consequence of
the *measurability* of the functional

$$\int_0^t V(x_0 + x(\tau)) \, d\tau.$$

The fact that one can relate solutions of differential equations to
function space integrals suggested many directions of further work.
By now the literature on the subject is so vast and so varied that even
a cursory review would be out of the question. In some instances ex-

---

[5] This amounts to constructing a Wiener like measure in the set of continuous
paths $y(\tau)$ such that $y(0)=0$ and $y(t)=y$. The simplest way to accomplish this is to
map the ordinary Wiener measure in space of paths $x(\tau)$ ($x(0)=0$) by means of the
mapping $y(\tau)=x(\tau)+(\tau/t)\,(y-x(t))$, $0 \leq \tau \leq t$.

[6] More precisely, the Wiener like measure described in the preceding footnote.

treme generality helps hide the true origins in Wiener's measure and integration, and one is reminded of a statement attributed (apocryphally perhaps) to Hilbert that the art of doing mathematics consists in finding that *special* case which contains all the germs of *generality*.

Feynman's ideas contributed greatly to gaining a larger audience for the work based on the Wiener integral and helped to take the "Wiener process" out of the somewhat narrow context of stochastic processes and into a much wider stream of analysis and physics.

As an illustration let me describe a simple application of the Wiener integral to quantum statistical mechanics.

The quantum-mechanical partition function is defined as the sum

$$(6.17) \qquad Z = \sum_{n=1}^{\infty} e^{-\beta E_n},$$

where the E_n's are the eigenvalues of the Schroedinger operator

$$-\frac{\hbar^2}{2m} D^2 + V(x) \qquad \left(D \equiv \frac{d}{dx} \right)$$

and

$$(6.18) \qquad \beta = \frac{1}{kT}$$

with T the absolute temperature and k the Boltzmann constant. It can be easily related to the so called Bloch equation

$$(6.19) \qquad \frac{\partial P}{\partial \beta} = \frac{\hbar^2}{2m} \frac{\partial^2 P}{\partial \beta^2} - V(x)P$$

which, except for units, is of the "real" heat equation (6.13). It was, of course, well known that

$$(6.20) \qquad Z = \int_{-\infty}^{\infty} P(x \mid x; \beta)\, dx$$

where $P(x_0 \mid x; \beta)$ is the Green's function of (6.19). But now, owing to (6.15), one could also write Z as a function space integral and in fact,

$$(6.21) \qquad
\begin{aligned}
Z = {}&\frac{m^{1/2}}{\hbar} \frac{1}{(2\pi\beta)^{1/2}} \\
&\cdot \int_{-\infty}^{\infty} E\left\{ \exp\left(-\int_0^{\beta} V\left(x + \frac{\hbar}{m^{1/2}} x(\tau) \right) d\tau \right) \middle| x(\beta) = 0 \right\} dx.
\end{aligned}$$

To appreciate the advantage resulting from the use of function space integrals consider the classical limit $h \to 0$.

It is now absolutely transparent what to expect.

Since as $h \to 0$

$$E\left\{\exp\left(-\int_0^\beta V\left(x + \frac{\hbar}{m^{1/2}}\, x(\tau)\right) d\tau\right) \Big| \, x(\beta) = 0\right\} \to \exp(-\beta V(x)),$$

we have formally

$$\lim_{h \to 0} (\hbar Z) = \frac{m^{1/2}}{(2\pi\beta)^{1/2}} \int_{-\infty}^{\infty} \exp(-\beta V(x))\, dx$$

$$= \int_{-\infty}^{\infty} \exp\left(-\beta\left(\frac{p^2}{2m}\right)\right) dp \int_{-\infty}^{\infty} \exp(-\beta V(x))\, dx$$

$$= \int_{-\infty}^{\infty} \int_{-\infty}^{\infty} \exp\left(-\beta\left(\frac{p^2}{2m} + V(x)\right)\right) dx\, dp$$

which is the classical partition function since

$$\frac{p^2}{2m} + V(x)$$

is the Hamiltonian of the system.

The formal steps can be justified at the expense of some mild conditions on V. Under appropriately stronger conditions on V (6.21) can be made a starting point (as was, in fact, done by Siegert and Yaglom) of a simple and straightforward derivation of the Kirkwood-Wigner expansion of Z in powers of h.

While on the subject of quantum mechanics one should at least mention another line of development of integration in function spaces. This has to do with an integration theory in Hilbert space and was dictated by certain needs of quantum theory of fields. Originated by Friedrichs and I. E. Segal this theory has, more recently, been shown to be closely related to and, in some respects, equivalent with Wiener's integration theory in the space of continuous functions. As one example of the interplay of the two "schools of thought" we may mention e.g. L. Gross' Hilbert space version of the striking inversion formula of Cameron and Donsker which they discovered and proved in the context of Wiener integrals.

Also the Cameron-Martin theory of linear changes of variable alluded to above assumes a particularly appealing form in Hilbert space setting.

7. Wiener did not directly participate in the evolution described above and in fact, the only reference to it in his autobiography [177] is the brief statement (pp. 179–180) that "He (Martin) and Cameron did a good deal of work together along the lines of my Brownian motion papers, and they organized the field into a generally recognized branch of mathematical work."

Like all highly creative and highly original people Wiener preferred thinking to reading and he did not take the trouble of familiarizing himself with the rapidly growing body of literature. But he had (and justly so!) a strong proprietory feeling toward his integral and he watched the work related to it closely enough to raise a few questions of priority.

What is perhaps really surprising is that he was apparently unaware of the intimate connections between his measure and his own work on potential theory.

It is now almost universally known that the generalized Wiener-Perron capacitory potential of a closed set F (in Euclidean space of dimension 3 or higher) at a point p is the Wiener measure of the set of paths which originate from p and which at some time hit F. Moreover, Wiener's famous criterion for regularity of boundary points has a most appealing interpretation in terms of his measure.[7]

The surprise is heightened if one recalls that Wiener's work on potential theory was almost simultaneous with that on Brownian motion.

But while Wiener did not take active part in most of the probabilistic and analytic developments which his early work inspired his interest in the preoccupation with random functions always remained strong.

I have already mentioned his book [191].

One should also mention his (somewhat obscurely motivated) paper on Homogeneous Chaos [108] which, more recently, in the hands of Ito, Kakutani and Segal, has found interesting uses in the theory of unitary operators and in tensor algebras over Hilbert spaces.

Finally, there was the attempt to use randon functions to achieve a sort of "hidden variables" interpretation of quantum mechanical probabilities. This was first discussed in the paper [167] written jointly with Armand Siegel and in spite of a rather indifferent reception it received, Wiener had high hopes for this approach. It was taken up again in [209] which is probably his last published paper.

---

[7] A very complete discussion of this subject can be found in the recent book of Ito and McKean *Diffusion processes and their sample paths*, Springer-Verlag, New York.

8. In retrospect one can have nothing but admiration for the vision which Wiener had shown when, almost half a century ago, he had chosen Brownian motion as a subject of study from the point of view of the theory of integration. To have foreseen, at that time, that an impressive edifice could be erected in such an esoteric corner of mathematics was a feat of intuition not easily equalled now or ever.

It was Josiah Willard Gibbs, whom Wiener admired so much who said that "one of the principal objects of the theoretical research in any department of knowledge is to find the point of view from which the subject appears in its greatest simplicity."

Integration in function spaces provided such a point of view over and over again in widely scattered areas of knowledge and it gave us not only a new way of looking at problems but actually a new way of *thinking* about them.

The fate of all great work is to be subsumed; the more attention it attracts the greater the chances of becoming engulfed in a cascade of generalizations and extensions.

This is especially true today because of a growing tendency to believe that the latest improvement supersedes all that preceded it and that a generalization constitutes a license to subsume.

It is therefore well to repeat that Wiener's contribution to the subject of integration in function spaces will forever be the greatest because he had the idea first; and should anyone try to attribute it to luck let him be reminded that it is the deserving ones who are also lucky.

THE ROCKEFELLER UNIVERSITY

WIENER'S WORK IN PROBABILITY THEORY

J. L. DOOB

In 1918 and 1919 Daniell's papers on what is now called the Daniell integral appeared. As an application he constructed the most general finite measure of Borel sets in Euclidean space of countable dimensionality. In 1933 Kolmogorov, in his formalization of probability as measure theory, rediscovered the Daniell result in constructing measures on Euclidean space of arbitrary dimensionality. After 1933 probability to a mathematician was no longer merely an extra-mathematical source of interesting problems in analysis with colorful interpretations, but was now a normal part of mathematics whose historical development was commemorated by the use of peculiar names (random variable, expectation, . . .) for commonplace mathematical concepts (measurable function, integral, . . .).

Wiener was an exception in his probability work in that he almost never used the classical nomenclature, and in fact usually even avoided using the standard classical results and conventions, some of which would have simplified and clarified his work. He came into probability from analysis and made no concession. If his work had been less forbiddingly formal, it might have had even more influence.

In a series of papers beginning with [12]* Wiener undertook a mathematical analysis of Brownian motion. It was accepted that Brownian paths were governed by probabilistic laws, and it seemed plausible that the paths were continuous. The problem was to construct and analyze a rigorous mathematical model. More than a decade before Kolmogorov's formalization of probability Wiener constructed a mathematical model of Brownian motion in which the basic probabilities were the values of a measure defined on subsets of a space of continuous functions. This measure has since been commonly called "Wiener measure." Fixing an origin in time and a direction in space, let $x(t)$ be the component in the specified direction of the displacement by time t of a Brownian particle. Then $x(0) = 0$. For technical reasons it was advantageous to restrict t to a compact interval. Thus Wiener was led to consider the space C of continuous functions on $[0, 1]$, vanishing at 0, and to define a measure of subsets of C (based on the Daniell integral). The probability of any property of the displacement function was associated with the measure of the subset of C having this property. The Wiener measure of subsets of C

---

* The bold-faced numbers in brackets refer to the numbered references in the Bibliography of Norbert Wiener.

has the following properties: (Here if ω is a member of C, $X(t, \omega)$ is the value of ω at t.)

(a) The measure of C is 1.

(b) The function $X(t, \cdot)$ is a random variable (measurable function) on C and $X(t_2, \cdot) - X(t_1, \cdot)$ has a distribution which is Gaussian with mean 0 and mean square value $\alpha |t_2 - t_1|$, where α is an assigned strictly positive parameter.

(c) If $t_0 < \cdots < t_n$, then $X(t_1, \cdot) - X(t_0, \cdot), \cdots, X(t_n, \cdot) - X(t_{n-1}, \cdot)$ are mutually independent.

In view of the definition of C the values of t here are restricted to the interval $[0, 1]$ but a simple transformation then yields a measure on the space of continuous functions on $[0, \infty)$, or $(-\infty, \infty)$ if desired, vanishing at 0, with properties (a), (b), (c) for the relevant parameter values. In his early work, and later also in his joint book $[92]$ with Paley, Wiener studied the regularity of Brownian paths, proving that almost no function in C is of bounded variation in any interval, and finding estimates of the moduli of continuity for the functions.

It is characteristic of his writing that in the details of his construction Wiener leaves it to the reader to recognize that property (c) is true, as an implication of the explicit distribution formulas he writes down. In general he preferred writing formulas to making descriptive remarks except in his later years. This preference helped the rigor but not the reader.

In $[92]$ $X(t, \omega)$ is defined in a different and very elegant way; $X(t, \cdot)$ is a function on the unit interval with Lebesgue measure, properties (a), (b), (c) hold, and $X(\cdot, \omega)$ is a continuous function for almost all ω. The construction does not involve the Daniell integral, and $X(t, \omega)$ is represented explicitly as the sum of a Fourier series with random coefficients.

In a little-known paper of 1921 $[18]$, Wiener applied his measure to obtain a second model for Brownian motion, a model which is the more exact model rediscovered by Ornstein and Uhlenbeck in 1930. In this model the paths are defined by functions which (almost all) have continuous first derivatives but these derivatives have infinite variation on every interval.

The stochastic integral $I(f) = \int f(t) d_t X(t, \omega)$ was one of Wiener's most fruitful ideas. (We shall write d for d_t below.) Here $X(t, \omega)$ is from the Brownian motion process, first model, and f is Lebesgue measurable and square integrable on $(-\infty, \infty)$, $f \in L_2(-\infty, \infty)$. The integral, a random variable, can be defined in a reasonable way even though the function $X(\cdot, \omega)$ has infinite variation on every interval, for almost all ω. The transformation $f \rightarrow I(f)$ is a linear L_2-norm preserving transformation from $L_2(-\infty, \infty)$ into the L_2 space of square integrable random variables. The stochastic integral, which was generalized later by Ito to allow integrands f which may depend on ω as well as t, was for Wiener and still remains a fundamental tool in a variety of contexts, only a few of which will be mentioned below.

In [92] it is shown that Wiener's harmonic analysis of functions on $(-\infty, \infty)$ is applicable to functions defined by the stochastic integral $\int f(x+iy+t)dX(t, \omega)$, where f is a function analytic in the strip $|y| < c$ and satisfies certain boundedness conditions. Moreover the asymptotic properties of the zeros of h are derived.

Suppose that $\{Y(t), -\infty < t < \infty\}$ is a stationary stochastic process in the sense that $Y(t)$ is a random variable and that the inner product $R(t) = (Y(t+s), \overline{Y}(s))$ does not depend on s. Then the function $R(\cdot)$, supposed continuous, is the Fourier transform of a measure, the spectral measure of the process, which dominates the harmonic analysis of the process. Wiener showed, and applied repeatedly, that the Brownian motion process acts as though the derivative $X'(t, \omega)$ process exists, is stationary, and has a constant spectral density. For example, if A is a suitably restricted linear operator the equation

$$(1) \qquad\qquad A(Y) = Z(t, \omega)$$

can be solved to give a stationary $Y(t)$ process depending on the $Z(t)$ process, supposed given and stationary. The spectral measure dF_2 of the $Y(t)$ process is obtained from that of the $Z(t)$ process, dF_1, by means of a function g determined by A:

$$(2) \qquad\qquad dF_2 = |g|^2\, dF_1.$$

In many contexts (1) is replaced by

$$(1') \qquad\qquad A(Y) = X'(t, \omega)$$

where $X'(t, \omega)$ is from the fictitious derived process of the Brownian motion process and the equation is given meaning by a formal integration. In this way the solution of $(1')$ finally appears as a stochastic integral defining a stationary process and (2) is replaced by $dF_2 = \text{const}\,|g|^2\, dt$ in accordance with Wiener's principle. Equation $(1')$ arises for example in noise problems in which the right side represents "white noise," a "Brownian driver" a source which to a first approximation produces power uniformly distributed over all frequencies.

The preceding theory has been stated in linear terms. Wiener also discussed analogous problems in a nonlinear context [108], [191] in which he used multiple stochastic integrals as a fundamental tool. The translational shift $X(t, \omega) \rightarrow X(t+s, \omega)$ defines a probability preserving transformation of Wiener measure (here the t-interval is $(-\infty, \infty)$ and the condition $X(0, \omega) = 0$ is not imposed because only differences like $X(t_2, \omega) - X(t_1, \omega)$ are involved). Hence a multiple stochastic integral

$$\int \cdots \int f(t - t_1, \cdots, t - t_n) \, dX(t_1, \omega) \cdots dX(t_n, \omega) = h(t, \omega)$$

defines a family of random variables which is strictly stationary in the sense that joint distributions of sets of the $h(t, \cdot)$ are invariant under the shift. Wiener showed [191] how to evaluate moments of such multiple stochastic integrals and make a spectral analysis of the stationary process so defined. Moreover [108] he showed how families of this type "polynomial chaoses" could be used to approximate very general stationary processes. Ergodic theory is a natural tool in this area.

In summary, and neglecting Wiener's work in prediction theory which will be discussed elsewhere, Wiener's most important contributions to probability theory were centered about the Brownian motion process, now sometimes called the "Wiener process." He constructed this process rigorously more than a decade before probabilists had made their subject respectable and he applied the process both inside and outside mathematics in many important problems.

University of Illinois

WIENER'S CONTRIBUTIONS TO GENERALIZED HARMONIC ANALYSIS, PREDICTION THEORY AND FILTER THEORY

BY P. MASANI

0. Prologue. The strong cohesive forces permeating the work of Norbert Wiener complicate the task of surveying his contributions to specific areas. Where is one to begin and where to end? In the realm of prediction, for instance, Wiener's book [TS][1] stands out as his first major contribution. But an important part of this book concerns the synthesis of predictors, for which as Kakutani remarked (32): "The theory of generalized harmonic analysis developed by the author some 20 years ago is exactly the right tool . . . ." Now the latter theory, given in the memoir [GHA] of 1930, was itself the culmination of researches begun in 1924, which were motivated by even earlier investigations in the theory of Brownian motion. It would seem that a thorough review of Wiener's work in prediction should start from about the year 1919 when he looked at the Charles River from his office at M.I.T. and began to wonder whether the Lebesgue integral was the right tool for the analysis of the undulating water surface. Such a review would be beyond the abilities of this writer, even if he were granted the necessary space.

In this review we shall first survey those aspects of Wiener's great memoir [GHA] which bear on his later work on prediction and filtering (I). We shall then describe briefly how the mathematical activity of the thirties influenced his thought (II). Next we shall discuss Wiener's general theory of nonlinear prediction (III). From this we shall turn to his many contributions to linear prediction and filtering theory (IV). Lastly we shall dwell on his theory of filters (V).

I. GENERALIZED HARMONIC ANALYSIS (1930)

1. White light: the need for generalizing harmonic analysis. The optical origins of Wiener's work are best expounded from the standpoint of the electromagnetic theory of light. At a fixed point r in a medium traversed by light the direction of propagation at instant t is given by the Poynting vector $P(t) = E(t) \times H(t)$, where $E(t)$, $H(t)$

---

[1] The bold-faced numbers in brackets refer to the numbered references in the Bibliography of Norbert Wiener. The bold-faced letters in brackets and numbers in parentheses refer to the References at the end of this article.

are the electric and magnetic vectors at r at instant t. The *optical vector* is identified with $E(t)$; hence the instantaneous plane of optical vibration is that spanned by $E(t)$ and $P(t)$. When the light rays in the vicinity of r have a direction z independent of t, we can prescribe $E(t)$ by its two components $E_x(t)$, $E_y(t)$ in (fixed) perpendicular directions x, y orthogonal to z. The light signal is called *monochromatic* if E_x, E_y are *sinusoidal functions* of the same frequency;[2] otherwise *polychromatic*. Newton's experiments on dispersion revealed the polychromatic nature of sunlight, then called *white light*.

What sort of harmonic analysis can we make of a white light signal f ($=E_x$, say)?[3] Classical theory offers but two alternatives:

(1) Assume that f is a generalized trigonometric polynomial or its limit:

$$f(t) = \sum_{k=-\infty}^{\infty} c_k e^{i\omega_k t}, \qquad t \in (-\infty, \infty).$$

This assumption entails that the spectrum is made up of a finite or countable number of sharp lines of intensities $|c_k|^2$ at the frequencies ω_k, $k =$ an integer. What is observed, however, is a continuous band of colors.

(2) Assume that $f \in L_2(-\infty, \infty)$. Then by the Fourier-Plancherel Theorem

$$f(t) = \frac{1}{\sqrt{(2\pi)}} \int_{-\infty}^{\infty} c(\omega) e^{i\omega t}\, dt \qquad \text{(in the l.i.m. sense)}.$$

We now have as desired a continuous spectrum of energy on the frequency-band $(-\infty, \infty)$. But now

$$\int_{t}^{t+h} |f(\tau)|^2 d\tau \to 0, \qquad \text{as } t \to \pm\,\infty \quad (h \text{ fixed}).$$

This means that the energy[4] emitted by the signal during a time-interval of fixed length h approaches 0 as the interval advances (or recedes) indefinitely on the time axis. This result conflicts with the

---

[2] The function f is called *sinusoidal*, if $f(t) = A \cos(\omega t + \alpha) = \bar{c}e^{-i\omega t} + ce^{i\omega t}$, $t \in (-\infty, \infty)$. ω is the (*circular*) *frequency*, that is angular velocity, and $2\pi/\omega$ is the *period*. If $g(t) = ce^{i\omega t}$, $t \in (-\infty, \infty)$, then g is often called a *complex sinusoidal function of frequency* ω.

[3] It will suffice to study the analysis of just one of the component functions E_x, E_y.

[4] From Maxwell's equations the flux of electromagnetic energy through a small surface $(\delta S)r$ is proportional to $|E(t)|^2$, cf. (**64**, p. 333, (7–62)). Thus apart from a constant factor, $|E_\alpha(t)|^2$ represents the energy of the α-component.

standard theory in which the sun is treated as a permanent reservoir of infinite energy.[5]

Neither assumption seems to explain the behavior of white light satisfactorily. Wiener felt that the difficulty stemmed from the limitations of the classical harmonic analysis. Around 1924 he began developing a "generalized" analysis that could cover signals f on $(-\infty, \infty)$, which are on the one hand so irregular that their spectra are not made up of sharp lines alone, and on the other so lastingly vigorous that $\int_t^{t+h} |f(\tau)|^2 d\tau \nrightarrow 0$, as $t \to \infty$. In this venture he benefited from the ideationally deep though logically unrigorous investigations of the physicists Rayleigh, A. Schuster and G. I. Taylor, and from the work of mathematicians such as Hahn, Bohr and Bochner.[6] His researches culminated in his great memoir [GHA] of 1930. In this review we shall dwell only on those parts of [GHA] which bear on Wiener's later work in filter and prediction theory.[7]

NOTATION. We shall not adhere strictly to the notations and conventions used in [GHA] but adopt a few used by Wiener in later writings. For instance, we shall insert the factor $1/\sqrt{(2\pi)}$ in taking Fourier and Fourier-Stieltjes transforms, and write

$$(1.1) \quad \hat{f}(x) = \frac{1}{\sqrt{(2\pi)}} \int_{-\infty}^{\infty} f(y)e^{-ixy}\, dy, \quad \check{f}(y) = \frac{1}{\sqrt{(2\pi)}} \int_{-\infty}^{\infty} f(x)e^{ixy}\, dx.$$

We shall also use the abbreviations "SP" for "stochastic process," and BMSP for "the Brownian motion stochastic process."

2. **The Wiener classes $\mathcal{S}$ and $\mathcal{S}'$.** As Wiener emphasized, the germs of generalized harmonic analysis were already in the work of Schuster, but a "radical recasting" of the latter was necessary [GHA, p. 118]. From this recasting emerged the *covariance function* ϕ of the signal f:

$$(2.1) \quad \phi(\tau) = \lim_{T \to \infty} \frac{1}{2T} \int_{-T}^{T} f(t+\tau)\bar{f}(t)dt, \qquad \tau \in (-\infty, \infty).$$

Accordingly Wiener considered the class $\mathcal{S}$ of all Borel measurable functions f on $(-\infty, \infty)$, for which ϕ is defined on $(-\infty, \infty)$, as well as the subclass $\mathcal{S}'$ of $f \in \mathcal{S}$ for which ϕ is continuous at 0, and thence on $(-\infty, \infty)$.[8] We can show that $\mathcal{S}$ is closed under the trans-

---

[5] This conception of the sun is of course untenable from a cosmological viewpoint, but it is a legitimate idealization for the study of short term optical phenomena.

[6] The limitations of classical harmonic analysis had been felt in pure mathematics as well, for example in Dirichlet series, and in the statistical analysis of time series. The reader should consult the bibliography in [GHA] for references to earlier work.

[7] Other aspects of [GHA] are perhaps discussed elsewhere in this number.

[8] The classes $\mathcal{S}$, $\mathcal{S}'$ were actually introduced somewhat later, e.g., in [FI].

76   P. MASANI

lation group $(U_t, -\infty < t < \infty)$, $(U_t f)(\tau) = f(t+\tau)$, and that f and $U_t f$ have the same ϕ, [FI, p. 155]. But $\mathcal{S}$ is not a vector space. Also, as is easy to check, ϕ is a positive definite function.

Wiener then introduced the function S of which ϕ is the Fourier-Stieltjes transform, calling it the *integrated spectrum* of f. (The current term is *spectral distribution* of f.) To appraise this work let us ignore order of precedence and proceed logically. Two years after the appearance of [GHA] came Bochner's extension to the real line of Herglotz's theorem on the Fourier-Stieltjes representability of continuous positive definite functions (5, §20). In 1939 Cramer (11) showed that the continuity assumption is dispensable. Now let $f \in \mathcal{S}$. Then since ϕ is positive definite, it follows from the Bochner-Cramer Theorem that

$$(2.2) \qquad \phi(t) = \frac{1}{\sqrt{(2\pi)}} \int_{-\infty}^{\infty} e^{-itu} dS(u), \quad \text{a.e.,}$$

where S is a bounded monotone increasing function, unique after proper normalization. It is easy to check that an admissible version of S is

$$(2.3) \qquad S(u) = S(0) + \lim_{A \to \infty} \frac{1}{\sqrt{(2\pi)}} \int_{-A}^{A} \phi(t) \frac{e^{iut} - 1}{it} dt.$$

Wiener was obliged to proceed in a different way. He defined S by

$$(2.4) \qquad S(u) = \underset{A \to \infty}{\text{l.i.m.}} \frac{1}{\sqrt{(2\pi)}} \left(\int_{-A}^{-1} + \int_{1}^{A} \right) \frac{\phi(t) e^{iut}}{it} dt$$
$$+ \frac{1}{\sqrt{(2\pi)}} \int_{-1}^{1} \phi(t) \frac{e^{iut} - 1}{it} dt,$$

[GHA, (3.19)–(3.21)][9] and showed by hard work that it could be redefined on a set of zero measure so as to make it monotonic increasing and in fact to satisfy (2.3), [GHA, (3.27), (3.28)]. He then showed that S is bounded; and indeed that

$$\frac{1}{\sqrt{(2\pi)}} \{ S(\infty) - S(-\infty) \} = \lim_{\epsilon \to 0} \frac{1}{2\epsilon} \int_{-\epsilon}^{\epsilon} \phi(t) dt$$

$$(2.5) \qquad\qquad\qquad \leq \phi(0) = \lim_{T \to \infty} \frac{1}{2T} \int_{-T}^{T} | f(t) |^2 dt,$$

---

[9] The factor $1/2\pi$ occurs in [GHA] in place of $1/\sqrt{(2\pi)}$. But in later works, e.g., [FI, TS] Wiener uses $1/\sqrt{(2\pi)}$. Our choice of e^{iut} (and not e^{-iut}) in (2.4) and (2.3) agrees with that made in [GHA] and for SP's in [185, 186]. It has the merit of yielding holomorphic functions in the upper (and not lower) half plane in cases of interest, but the disadvantage that the power distribution of f is not $S/\sqrt{(2\pi)}$ but rather its "reflection" about the origin, cf. (4.4) et seq. below.

and that equality prevails if and only if $f \in \mathsf{s}'$ [**GHA**, (4.09), (4.10), (5.46)]. Only then was Wiener able to recover ϕ from S by the equation (2.2), [**GHA**, (5.40)].

We see that Wiener's treatment of S would have been appreciably simpler had the Bochner-Cramer results been at his disposal. But even in later writings he did not appeal to Bochner's theorem to define S but continued to use (2.4), cf. e.g., [**TS**, p. 42]. He did this perhaps to keep in sight the formal similarity between S and the important function s (below) which he had to define as a l.i.m.

The Wiener theory of S received an important generalization when Bochner defined S for the wider class of functions f such that

$$(2.6) \qquad \lim_{T \to \infty} \frac{1}{\rho(T)} \int_{-T}^{T} f(t + \tau)\bar{f}(\tau)dt$$

exists for all t in $(-\infty, \infty)$, where ρ is any positive increasing function on $(0, \infty)$ for which $\rho(T+1)/\rho(T) \to 1$, as $T \to \infty$, (5, pp. 294–295).

We must next turn to another important function which Wiener associates with a function f in s, viz. its *integrated* (or *generalized*) *Fourier transform* s. On comparing the s-theory with the classical L_2-theory we find that the covariance function ϕ is the natural counterpart of the inner product function ϕ^* in L_2-theory defined by

$$\phi^*(t) = \int_{-\infty}^{\infty} f(t + \tau)\bar{f}(t)dt = \frac{1}{\sqrt{(2\pi)}} \int_{-\infty}^{\infty} e^{-itu} \,|\, s^*(u) \,|^2 du$$

where $s^* = \bar{f}$, (cf. (1.1)). Since this can be rewritten

$$\phi^*(t) = \frac{1}{\sqrt{(2\pi)}} \int_{-\infty}^{\infty} e^{-itu} dS^*(u), \quad \text{where } S^*(u) = \int_{-\infty}^{u} |\, s^*(\lambda) \,|^2 d\lambda,$$

we see that Wiener's S is the counterpart of S^*. *Does the Fourier-Plancherel transform s^* itself have a counterpart s in the s-theory?* If so, s would give the "generalized harmonic analysis" of an f in s in somewhat the same way that s^*, i.e. $\bar{f}$, gives the "harmonic analysis" of an f in L_2. Wiener gave an affirmative answer to this question and established for s generalizations of classical results for $\bar{f}$. In doing so he has left us with some hard analysis for which no simpler substitute seems to have been found during the last 35 years.

Wiener's work on s resembles to some extent that of Kolmogorov and Cramer on the spectral representation of a weakly stationary SP, a topic viewed best from the standpoint of Stone's spectral resolution for unitary flows on Hilbert spaces. The spectral representation involves stochastic integrals with values in the Hilbert space $L_2(\Omega, \mathcal{B}, P)$. One might attempt to retrieve Wiener's results on s from this representation by plugging in the random parameter ω. But many of these results pertain to the time-averages of (very irregular) functions in $\mathcal{S}$, and it is doubtful if any except the very trivial can be had in this way, cf. Doob (**19**).

We have thus to revert to Wiener's hard analysis involving "certain weighted averages" the appropriate tool for handling which is the "general theory of the Tauberian theorems developed by the author and applied to these problems by Mr. S. B. Littauer" [**GHA**, p. 119]. In this review we cannot go into this question. It will suffice to say that Wiener defines s by the equation obtained from (2.4) by replacing ϕ by f, and to mention two very interesting results concerning s. First,

$$(2.7) \quad \lim_{\epsilon \to 0} \frac{1}{4\pi\epsilon} \int_{-\infty}^{\infty} |s(u+\epsilon) - s(u-\epsilon)|^2 dx = \lim_{T \to \infty} \frac{1}{2T} \int_{-T}^{T} |f(t)|^2 dt;$$

in words, *the "quadratic variation" of s equals the "mean-square modulus" of f* [**GHA**, (5.53) & p. 119], [**TS**, (20.13)]. Next, [**TS**, p. 160, Theorem 28] an $f \in \mathcal{S}$ will be in $\mathcal{S}'$ if and only if

$$(2.8) \quad \lim_{A \to 0} \lim_{\epsilon \to 0} \sup \frac{1}{4\pi\epsilon} \left(\int_{-\infty}^{-A} + \int_{A}^{\infty} \right) |s(u+\epsilon) - s(u-\epsilon)|^2 du = 0.$$

We should also point out the limitations of s. Whereas the spectral representation involves a stochastic integral, Wiener's corresponding equation [**GHA**, (6.01)]:

$$f(t) = \frac{1}{\sqrt{(2\pi)}} \int_{-\infty}^{\infty} e^{-itu} \, ds(u), \quad \text{a.e.,}$$

makes sense only when the right-hand side is interpreted as the Cesaro limit of a pseudo-Stieltjes integral, i.e., one defined by a formal integration by parts. This is so because s is not of bounded variation except in trivial cases, [**GHA**, (6.02), (6.05)]. Let us not forget, however, that the *stochastic integral* which Cramer and others have used with much success, but which Wiener could not use in his $\mathcal{S}$-theory because of the absence of a random parameter, first appeared in Wiener's own work [**29**] of 1923, the integrator process of orthogonal increments being the BMSP. Let us add that Wiener also defined s for a wider class of functions than $\mathcal{S}$, viz. the set of f for which $(1/2T)\int_{-T}^{T}|f(t)|^2 dt$ is bounded, and the subset of the Bochner class for which the function ρ in (2.6) is $\rho(T) = T^n$, n being an odd positive integer [**GHA**, §6].

3. **Multiple harmonic analysis and coherency.** Wiener was able to

extend the foregoing ideas to vectorial signals with two or more components $f_1, \cdots, f_q \in \mathcal{S}$, [**GHA**, §9]. We assume that for $1 < i, j < q$,

$$(3.1) \qquad \phi_{ij}(\tau) = \lim_{T \to \infty} \frac{1}{2T} \int_{-T}^{T} f_i(t + \tau)\overline{f_j(t)}dt \quad \text{exists.}$$

Then by the Schwarz inequality

$$|\phi_{ij}(\tau)|^2 \leqq \phi_{ii}(0)\phi_{jj}(0).$$

We are thus led to the $q \times q$ matrix-valued function $\phi = [\phi_{ij}]$, called the *covariance function* of the q-ple signal $f = (f_1, \cdots, f_q)$. ϕ_{ij} is called the *covariance* of f_i and f_j.

Again let us discard the historical order and proceed logically. The equations (3.1) reduce to the single matrix equation

$$(3.2) \qquad \phi(\tau) = \lim_{T \to \infty} \frac{1}{2T} \int_{-T}^{T} f(t + \tau)f^*(t)dt$$

in which f is a column vector and f^* is its adjoint and therefore a row vector. The integrand is thus the $q \times q$ matrix with

$$f_i(t + \tau)\overline{f_j(t)}$$

in the (i, j)th entry. From this it readily follows that ϕ is a *function of non-negative type*, i.e.,

$$\phi(-t) = \phi(t)^*,$$

and for all $q \times q$ matrices $C_1, \cdots, C_n$ and all real numbers $t_1, \cdots, t_n$,

$$\sum_{i=1}^{n} \sum_{j=1}^{n} C_i \phi(t_i - t_j) C_j^* - \text{\textit{a non-negative hermitian matrix.}}$$

Hence by the matricial extension of Bochner's Theorem, cf. Cramer (11), (12)[10] there exists a bounded $q \times q$ hermitian matrix-valued function $S = [S_{ij}]$ on $(-\infty, \infty)$ with non-negative increments such that

$$(3.3) \qquad \phi(t) = \frac{1}{\sqrt{(2\pi)}} \int_{-\infty}^{\infty} e^{-it\lambda} dS(\lambda), \text{ a.e.}$$

i.e.,

$$\phi_{jk}(t) = \frac{1}{\sqrt{(2\pi)}} \int_{-\infty}^{\infty} e^{-it\lambda} dS_{jk}(\lambda), \text{ a.e.}, \qquad 1 \leqq j, k \leqq q.$$

---

[10] Our concept of *matricial non-negative type functions* is not delineated in Cramer's papers [11], [12]. But by introducing this concept we can get the matricial extension of Bochner's Theorem by using the very arguments which Cramer advances

Since each S_{jk} is of bounded variation and S_{jj} is real and monotonic increasing, these equations make sense. Again Wiener was obliged to proceed differently. He defined S_{jk} by

$$S_{jk}(u) = \frac{1}{\sqrt{(2\pi)}} \int_{-\infty}^{\infty} \phi_{jk}(t) \, \frac{e^{iut} - 1}{it} \, dt,$$

asserting that the last integral converges and that the matrix $[S_{ij}]$ so obtained has the properties just mentioned. (There is unfortunately some hand waving in this section of [GHA].)

The matrix function S is called nowadays the *spectral distribution function* of the vectorial signal f. Wiener called it the *coherency matrix* of f. It is trivial to check that if $g(t) = Af(t)$, where A is a constant $p \times q$ matrix, then the covariance and spectral matrices of the p-ple signal g are $A\phi A^*$ and ASA^*. In particular taking $p=1$ we get the covariance and coherency matrices of the signal $f = \sum_{j=1}^{q} a_j f_j$:

$$(3.4) \qquad \phi(t) = \sum_{j=1}^{q} \sum_{k=1}^{q} a_j \bar{a}_k \phi_{jk}(t), \qquad S(\lambda) = \sum_{j=1}^{q} \sum_{k=1}^{q} a_j \bar{a}_k S_{jk}(\lambda).$$

Wiener defined two signals $f_1, f_2 \in \mathcal{S}$ to be *incoherent*[11] when the cross-covariance function ϕ_{12} or (equivalently) the cross spectral distribution S_{12} vanishes identically, i.e., when the covariance and spectral matrices of the 2-ple signal $f = (f_1, f_2)$ are diagonal:

$$\phi = \begin{bmatrix} \phi_{11} & 0 \\ 0 & \phi_{22} \end{bmatrix}, \qquad S = \begin{bmatrix} S_{11} & 0 \\ 0 & S_{22} \end{bmatrix}.$$

When the signals $f_1, \cdots, f_q$ are pairwise incoherent, (3.4) reduces to

$$\phi(t) = \sum_{j=1}^{q} |a_j|^2 \phi_{jj}(t), \qquad S(\lambda) = \sum_{j=1}^{q} |a_j|^2 S_{jj}(\lambda).$$

4. **Optical power, coherence and polarization.** Wiener's theory of the $\mathcal{S}$ class helped to clarify the optical concepts of power (i.e. intensity or brightness), coherence, and polarization. The statistical character of these concepts had been discerned by the physicists but they were unable to unravel the mathematical intricacies. The following account is based on Wiener's writings in optics [66], [168] and not just on [GHA] where the relevant exposition is somewhat sketchy.

(i) In view of the proportionality of $|E(r, t)|^2$ to the flux of electromagnetic energy through a small surface at r perpendicular to the direction of propagation (cf. Footnote 4), it is natural to define

---

[11] He did not do so explicitly in [GHA], but his remarks suggest that he had it in mind, cf. [TS, p. 45].

$$(4.1) \qquad \textit{the total energy of } f = \int_{-\infty}^{\infty} |f(t)|^2 dt,$$

$$(4.2) \qquad \textit{the total power}^{12} \textit{ of } f = \lim_{T \to \infty} \frac{1}{2T} \int_{-T}^{T} |f(t)|^2 dt = \phi(0).$$

Thus all signals f in $\mathcal{S}$ have finite power and the nontrivial ones (for which $\phi(0) \neq 0$), such as steady white light signals, have nonzero power and infinite energy. Signals in $L_2(-\infty, \infty)$, so-called *transients* or *pulses*, have zero power and finite energy.

For the complex sinusoidal signal, $f(t) = ce^{i\omega t}$, $t \in (-\infty, \infty)$, it follows at once from (4.2) that the total power is $|c|^2$. Next for the signal

$$(4.3) \qquad f(t) = \sum_{k=1}^{n} c_k e^{i\omega_k t}, \qquad t \in (-\infty, \infty)$$

we easily find that

$$(4.4) \qquad \phi(t) = \sum_{k=1}^{n} |c_k|^2 e^{i\omega_k t}, \qquad S(u)/\sqrt{(2\pi)} = \sum_{-\infty < -\omega_k \leq u} |c_k|^2.$$

Thus $\{ S(u_2) - S(u_1) \}/\sqrt{(2\pi)}$ is the power due to the frequencies in the interval $[-u_2, -u_1)$. In other words, for the signal f in (4.3) *the reflection about O of $S/\sqrt{(2\pi)}$ gives the distribution of power or intensity over the frequency band.* Naturally, we retain this interpretation of S when f is any arbitrary signal in $\mathcal{S}$. For signals $f \in \mathcal{S}'$,

$$\{ S(\infty) - S(-\infty) \}/\sqrt{(2\pi)}$$

equals $\phi(0)$, the total power of the signal f, cf. (2.5) et seq.; but for signals $f \in \mathcal{S} - \mathcal{S}'$, $\{ S(\infty) - S(-\infty) \}/\sqrt{(2\pi)}$ is the power due to *finite* frequencies only, and $\phi(0) - \{ S(\infty) - S(-\infty) \}/\sqrt{(2\pi)}$ is that arising from the frequencies $\pm \infty$. Thus *the class $\mathcal{S} - \mathcal{S}'$ comprises precisely those signals which have infinite frequencies in addition to finite frequencies.*

(ii) In the preliminary stages of optics light signals f_1, f_2 are considered to be coherent or interfering when the intensity of the signal $f_1 + f_2$ is "noticeably greater" or "noticeably smaller" than that "ordinarily" observed. The signals are said to be incoherent when the phase difference "varies rapidly and irregularly with time," so that to the eye or other optical device the result appears as a steady intensity, cf. Rossi (**64**, p. 119). Wiener explicated this provisional and imprecise concept of coherence by defining f_1, f_2 to be *completely incoherent* when the cross-covariance $\phi_{12}(t) = 0$ for all $t \in (-\infty, \infty)$, and to be *completely coherent* in case $\phi_{12}(t) = \phi_{11}(t)$ for each t. Following him,

---

[12] In [GHA] Wiener uses the term "energy," but this is rightly discarded in favor of "power" in later writings, e.g. in [TS, p. 42].

we may interpret ϕ_{12} or a suitable function containing it, such as $\phi_{12}/\sqrt{(\phi_{11}\phi_{22})}$ or the matrices ϕ and S, as *measures of coherence.*

With these definitions Wiener was able to explain why the intensity I of the superposition f_1+f_2 of two completely incoherent signals f_1, f_2 is I_1+I_2. This fact[13] was puzzling, since Maxwell's equations tell us that the intensity is proportional to $|f(t)|^2$, and obviously $|f_1(t)+f_2(t)|^2 \neq |f_1(t)|^2+|f_2(t)|^2$. Wiener's explanation went somewhat as follows. Since neither the eye nor any other optical device can resolve the very rapid fluctuations of intensity which occur because of the very large number of independent atomic sources involved, the observed intensity of f is really a time-average $I=(1/2T)\int_{-T}^{T}|f(t)|^2dt$. In this average even a short duration T, e.g. 1 microsecond, is so great from the atomic standpoint that the error committed in letting $T \to \infty$ and replacing I by $\phi(0)$ is insignificant. In effect, *the observed intensity of f is $\phi(0)$.* Now in case $f=f_1+f_2$ we have

$$\phi(0) \;=\; \phi_{11}(0) \;+\; \phi_{22}(0) \;+\; 2 \text{ real } \phi_{12}(0)$$

which, when the light signals are incoherent, i.e. $\phi_{12}(0)=0$, reduces to $I=I_1+I_2$.

(iii) A light signal with a fixed direction of propagation z at a point r is called *elliptically polarized,* if the tip of $E(t)$ describes an ellipse in the plane $\pi \perp z$. (Circularly and plane polarized lights are obvious special cases.) The signal is said to be *unpolarized* when the tip of $E(t)$ moves "rapidly and erratically" in π. To explicate this interesting but imprecise idea Wiener took an orthonormal basis $B = \{x, y\}$ in π, and represented the light signal relative to B by the 2-ple signal $f_B = (E_x, E_y)$. f_B will of course have a 2×2 matricial covariance function ϕ_B. Wiener defined the light signal to be *completely unpolarized* if for some basis B in π,

$$(4.5) \qquad\qquad \phi_B(t) \;=\; \phi(t)I, \qquad t \in (-\infty, \infty).$$

For any other orthonormal basis B' in π we have obviously

$$\phi_{B'}(t) \;=\; U\phi_B(t)U^*$$

where U is an orthogonal matrix. Hence light is completely unpolarized, if and only if the equation (4.5) holds for all orthonormal bases B in π. Obviously the spectral distribution (or coherency) matrix for such a signal will take the form $S(\cdot)I$ in any basis B.

On the other hand, for elliptically polarized monochromatic light of frequency ω, we find that relative to the orthonormal basis formed by the principal axes of the ellipse

---

[13] Crudely expressed by the statement "two candles are twice as bright as one."

$$f_B(t) = (a \cos (\omega t + \alpha), \, b \sin(\omega t + \alpha)),$$

$$\phi_B(t) = \frac{1}{2} \begin{bmatrix} a^2 \cos \omega t & -ab \sin \omega t \\ ab \sin \omega t & b^2 \cos \omega t \end{bmatrix}.$$

This ϕ_B cannot be diagonalised by a (real) rotation of coordinate axes. Wiener proved that any partially polarized monochromatic signal is decomposable into a completely unpolarized signal and an elliptically polarized one. He also discussed the idea of "degree of polarization" and the structure of the group of optical transformations [GHA, §9].

Wiener's ideas have found a place in standard works on optics, e.g. in Chapter X in *Principles of optics* by M. Born and J. Wolf. See also §2 in the paper *Fluctuations of light beams* by L. Mandel in Progress in Optics, Vol. II. His ideas appear in such works in a garb which may seem strange to mathematical readers.

5. Signals with absolutely continuous spectra defined by random processes. In [GHA, §§11–13] Wiener gave examples to show that signals in $\mathbb{S}'$ can have spectra of all possible types: saltus, singular, or absolutely continuous. Two examples for the last type are derived from stochastic processes, and are especially interesting in pointing to later developments. In our comments we have simplified Wiener's discussion.

EXAMPLE 1 [GHA, §12]. Let $\tilde{r}_k$ be the kth Rademacher function on $[0, 1]$.[14] After Borel we know that the $\tilde{r}_k$ form an independent family of variates on $([0, 1], \mathfrak{B}, \text{Leb.})$, where $\mathfrak{B}$ is the family of Borel subsets of $[0, 1]$. Consequently the bisequence

$$(r_n)_{n=-\infty}^{\infty} = (\cdots, \tilde{r}_6, \tilde{r}_4, \tilde{r}_2, \tilde{r}_1, \tilde{r}_3, \tilde{r}_5, \cdots)$$

where $r_n = \tilde{r}_{2(1-n)}$ or $\tilde{r}_{2n-1}$ according as $n \leq 0$ or $n \geq 1$, is an independent, stationary, discrete parameter SP. In essence, Wiener built from this a continuous parameter SP $x(\cdot, \cdot)$ over $([0, 1], \mathfrak{B}, \text{Leb.})$, viz.

$$(5.1) \quad x(t, \alpha) = \sum_{k=-\infty}^{\infty} \chi_{J_k}(t) r_k(\alpha), \qquad t \in (-\infty, \infty), \quad \alpha \in [0, 1]$$

where $J_k = [k-1/2, \, k+1/2)$.[15] He showed that for almost all α, the signal or path $x(\cdot, \alpha)$ is in $\mathbb{S}'$, $\phi_\alpha = \phi$ and $S_\alpha = S$, where ϕ is the triangular function

---

[14] I.e., $\tilde{r}_k(\alpha) = 1 - 2d_k(\alpha)$, $0 \leq \alpha \leq 1$, $k \geq 1$, where $d_k(\alpha)$ is the kth digit in the binary expansion of α.

[15] χ_J denotes the indicator function of J. Actually Wiener took $J_k = [k, \, k+1)$, but our choice of J_k is a little nicer.

$$\phi(t) = \begin{cases} 1 - |t|, & -1 \leq t \leq 1 \\ 0, & |t| > 1 \end{cases}$$

and S is absolutely continuous on $(-\infty, \infty)$ with

$$(5.2) \qquad S'(u) = \sqrt{(2\pi)} \, |\tilde{\chi}_{J_0}(u)|^2 = \sqrt{\left(\frac{2}{\pi}\right)} \, \frac{1 - \cos u}{u^2}.$$

To understand this example better, let $\xi(\Delta) = \sum_{k \in \Delta} r_k$ for any Borel subset Δ of $(-\infty, \infty)$. Then ξ is a measure with values in $L_2[0, 1]$ such that $\xi(\Delta_1)$, $\xi(\Delta_2)$ are independent when Δ_1, Δ_2 are disjoint—a so-called *countably additive, independently scattered (c.a.i.s.) measure*. In terms of ξ we can rewrite (5.1) as an equation with stochastic integrals:

$$(5.3) \qquad x(t, \cdot) = \int_{-\infty}^{\infty} \chi_{[\tau-1/2, \tau+1/2)}(t)\xi(d\tau) = \int_{-\infty}^{\infty} \chi_{J_0}(t - \tau)\xi(d\tau).$$

The process $x(\cdot, \cdot)$ is thus a (2-sided) moving average of the c.a.i.s. measure ξ. But since ξ is concentrated on the set of integers, its variance-measure $|\xi(\cdot)|_2^2$ is not invariant under the translation group on $(-\infty, \infty)$. Hence the process $x(\cdot, \cdot)$ is not stationary. Thus Wiener has exhibited *a nonstationary SP almost all signals (i.e. sample-functions) of which are "stationary"* in the sense of belonging to $\mathcal{S}'$ and having the same covariance and spectral functions ϕ and S.

EXAMPLE 2 [**GHA**, §13]. The last example shows that the signal resulting from "a haphazard sequence of positive and negative rectangular impulses" almost always has the spectral density S' given in equation (5.2), but that such a signal cannot be derived from a stationary SP. With remarkable insight Wiener saw that to remedy this situation one has "to eliminate the equal spacing of the individual impulses, to reduce the sequence of impulses to such an irregularity as is found in the Brownian motion" (p. 213). To put it in contemporary terms, we must replace the discrete c.a.i.s. measure ξ of Example 1 by the Brownian measure η with values in $L_2[0, 1]$ (built from the increments of the BMSP) for which the spectrum is the entire real line $\mathcal{R}$ and the variance measure $|\eta(\cdot)|_2^2$ is Lebesgue and hence translation invariant.

The replacement of ξ by η led Wiener to the consideration of the SP $y(\cdot, \cdot)$ over $([0, 1], \mathcal{B}, \text{Leb.})$ defined by[16]

---

[16] We are unable to adhere to Wiener's notation. Our W is his ϑ (p. 225), our η is the measure (set-function) induced by his point-function ξ. Our $y(\cdot, \cdot)$ is his f in (13.36). His $\xi(t)$, $f(t)$ should of course be written $\xi(t, \alpha)$, $f(t, \alpha)$, with $t \in (-\infty, \infty)$,

$$(5.4) \qquad y(t, \cdot) = \int_{-\infty}^{\infty} W(t - \tau)\eta(d\tau), \qquad W \in L_2(-\infty, \infty).$$

By $5\frac{1}{2}$ pages (pp. 226–231) of hard analysis in which a generalization of the Tauberian result (2.7) is used, he was able to show that

$$(5.5) \qquad \begin{aligned} &\textit{almost every realization } y(\cdot, \alpha) \textit{ of the process } y(\cdot, \cdot) \textit{ belongs to}\\ &\textit{S', and } S_\alpha \textit{ is absolutely continuous on } (-\infty, \infty) \textit{ with } S_\alpha'\\ &= \sqrt{(2\pi)}|\tilde{W}(\cdot)|^2. \end{aligned}$$

The significance of this result and a short proof for it became apparent only after Birkhoff proved the ergodic theorem and Paley and Wiener [92, §40] showed that there is a flow T_t, $t \in \mathcal{R}$ such that

$$(5.6) \qquad \{\eta(\Delta)\}(T_t\alpha) = [\eta(\Delta + \{t\})](\alpha), \qquad \Delta \subseteq \mathcal{R}, \quad \alpha \in [0, 1],$$

and T_t is measure preserving and ergodic on $([0, 1], \mathcal{B}, \text{Leb.})$. T_t is referred to as *the flow of Brownian motion or of white noise*. Since it is Lebesgue measure-preserving, the process $y(\cdot, \cdot)$ is a strictly stationary moving average; in fact

$$(5.7) \qquad y(t + \tau, \alpha) = y(t, T_\tau\alpha).$$

Hence its covariance function γ and (absolutely continuous) spectral distribution F are given by

$$(5.8) \quad \gamma(t) = \int_{-\infty}^{\infty} W(t - \tau)\overline{W}(-\tau)d\tau, \qquad F'(\lambda) = \sqrt{(2\pi)}|\tilde{W}(\lambda)|^2,$$

cf. Doob (20, p. 532 ff). Now consider a realization $y(\cdot, \alpha)$ of the process $y(\cdot, \cdot)$. Since the T_t-flow is ergodic, therefore by Birkhoff's Theorem for almost all α

$$\begin{aligned} \phi_\alpha(t) &= \lim_{T \to \infty} \frac{1}{2T} \int_{-T}^{T} y(t + \tau, \alpha) \cdot y(\tau, \alpha)d\tau\\ &= \lim_{T \to \infty} \frac{1}{2T} \int_{-T}^{T} y(t, T_\tau\alpha) \cdot y(0, T_\tau\alpha)d\tau \qquad \text{by (5.7)}\\ &= \int_{0}^{1} y(t, \alpha) \cdot y(0, \alpha)d\alpha = \gamma(t). \end{aligned}$$

Hence by (5.8) for almost all α, $S_\alpha' = F' = \sqrt{(2\pi)}|\tilde{W}(\cdot)|^2$.

$\alpha \in [0, 1]$. Wiener uses ξ, f to refer sometimes to the random variable and sometimes to the signal. In place of our (5.4) he has (13.36)

$$y(t, \alpha) = \int_{-\infty}^{\infty} \xi(\tau, \alpha)dW(t + \tau).$$

With his additional assumptions on W this equation reduces essentially to (5.4) on integrating by parts. But the last integral has filter theoretic significance, cf. §23 below.

We see from this that although the flow T_t is not mentioned in [GHA], Wiener's $5\frac{1}{2}$ page proof of (5.5) was in effect a direct proof of the existence (a.e.) of time averages for this flow—a remarkable analytic feat.

II. Assimilation and consolidation (1930–1940)

Wiener did some of his best mathematical work during the thirties. But as far as research in prediction and filter theory is concerned, the period was primarily a preparatory one in which he consolidated his intellectual position. Several strands are discernible in the mathematical activity of this period, which had a profound effect on Wiener's later work, and which therefore call for comment. As before we shall eschew strict chronology.

6. Theory of stochastic processes. First we must mention the epoch-making contribution of Kolmogorov (37) in setting up the theory of stochastic processes on a sound footing. The basic concepts of random variate, conditional probabilities and expectations were cleared up. Wiener used these ideas later in his nonlinear prediction. Kolmogorov's treatment rested on the construction of a probability measure in an infinite dimensional product space, starting from a properly indexed hierarchy of marginally related probability spaces— a "stochastic family" in the sense of Bochner (6). A precursor of this general *Kolmogorov measure* was of course the *Wiener measure* in the space of continuous functions. Wiener was the pioneer, but the Kolmogorov systematization was a boon to all workers in the field.

Next came the formalization and study of the concept of a *stationary* SP due to Khinchine, Cramer, and others. We have already referred to the work of Bochner and Cramer on the Fourier-Stieltjes representability of continuous positive definite functions on $(-\infty, \infty)$ (§2). In 1934 Khinchine (36) showed that a weakly stationary SP has a spectral distribution function, and around 1940 Kolmogorov, Cramer and Loeve (38), (12), (44) showed that such a process admits a "spectral representation." Discrete-parameter and multivariate extensions of these results were made by H. Wold (66) and Cramer (12). All these contributions were to play an important part in Wiener's later work.

We must next refer to von Neumann's spectral theorem for unitary operators on Hilbert space and Stone's celebrated extension of this to one-parameter unitary groups. As Kolmogorov (38), (40) observed around 1940 these theorems provide an extremely elegant and unified treatment of the spectral theory of weakly stationary SP's. Kolmogorov and Karhunen (35) showed that associated with every *weakly* stationary SP $(f_t, -\infty < t < \infty)$ is a unitary group $(U_t,$

$-\infty <t< \infty$) such that $f_t = U_t(f_0)$. Here t runs over the set of integers or real numbers according as time is discrete or continuous. Assume that time is continuous, and let E be the spectral measure of the unitary group:

$$U_t = \int_{-\infty}^{\infty} e^{-it\lambda} E(d\lambda).$$

Then[17]

$$F(\lambda) = \sqrt{(2\pi)} \, | \, E(-\infty, \lambda] f_0 |^2, \qquad \xi_\lambda = E(-\infty, \lambda) f_0$$

give respectively the spectral distribution F of $(f_t, -\infty <t< \infty)$ and the associated process $(\xi_\lambda, -\infty <\lambda< \infty)$ of orthogonal increments; thus

$$(f_{t+h}, f_t) = \frac{1}{\sqrt{(2\pi)}} \int_{-\infty}^{\infty} e^{-it\lambda} dF(\lambda), \qquad f_t = \int_{-\infty}^{\infty} e^{-it\lambda} d\xi_\lambda.$$

Similar expressions are available in the discrete case. The results of Khinchine and Cramer thus become easy corollaries of Stone's Theorem, and the study of stationary SP's is reduced to that of *stationary curves* or *stationary sequences in Hilbert space.*

As Kakutani (32) remarks Wiener did not immediately avail himself of these immense simplifications. [TS] would have been easier on mathematical readers had he done so, though perhaps to engineers it might have been an even greater "yellow peril." However, Wiener adopted the Hilbertian approach in his later papers under the stimulus of his younger collaborators.

7. **Ergodic theory.** Stone's Theorem was destined to influence Wiener's work in yet another way. It suggested to Koopman the possibility of studying the asymptotic behavior of dynamical systems, governed by measure-preserving flows on a phase space (Ω, $\mathcal{B}$, P), by means of the spectral resolution and infinitesimal generator of the induced unitary group on $L_2(\Omega, \mathcal{B}, P)$. This led to the proof of the ergodic theorems of von Neumann, Birkhoff, and others in 1931.

It seems that at first Wiener regarded these theorems merely as useful tools to deal with his time averages. But under the influence of E. Hopf and by the natural evolution of his own thought he soon came to regard the Birkhoff theorem as a mighty beacon which made possible the rigorous construction of statistical mechanics as envisaged by J. W. Gibbs. The fact that ergodicity has to be postulated in order to go from time-averages to the expectations and other well-known averages of probability theory never bothered Wiener, for his

---

[17] The $\sqrt{(2\pi)}$ in the expression for F is in keeping with the conventions adopted for the spectra of signals in I above. In [185], [186] 2π is used instead of $\sqrt{(2\pi)}$.

scientific philosophy permitted the free creation of bold and ideal hypotheses. In 1938–1939 Wiener gave a unified treatment of different versions of the ergodic theorem and extended them to abelian groups with several generators [117], [108], [128].

8. **Hardy class functions.** In 1934 Paley and Wiener proved that $f \in L_2(-\infty, \infty)$ is the boundary value of a function f_+ in the Hardy class H_2 on the upper half plane, if and only if its direct Fourier-Plancherel transform $\hat{f}$ vanishes on $(-\infty, 0)$, [92, p. 8]. They also showed that the n.a.s.c. that a function $\phi \geq 0$ a.e. on $(-\infty, \infty)$ and in $L_1(-\infty, \infty)$ be expressible in the form $\phi = |f|^2$ a.e., where f is as just described, is that $\{\log \phi(\lambda)\}/(1+\lambda^2) \in L_1(-\infty, \infty)$, [92, pp. 16–17]. Both results play a central role in the theory of filters. In 1935 came R. Nevanlinna's book (57), in which the earlier work of F. Riesz, Nevanlinna, and Szegö on the canonical factorization of functions in the Hardy classes on the disc appeared as an elegant and coherent theory. Initially Wiener did not feel much need for this powerful theory in his work on prediction. But as he delved deeper into the field, he had to appeal to it.

9. **The Hopf-Wiener integral equation.** Wiener's interest in integral equations in which the integral is a one-sided convolution was aroused by his colleague E. Hopf. The general H.W. equation of the second kind with unknown f is

$$f(t) + \int_0^\infty K(t - x)f(x)dx = g(t), \qquad t \geq 0,$$

the corresponding equation of the first kind (encountered in prediction and filtering) being

$$\int_0^\infty K(t - x)f(x)dx = g(t), \qquad t \geq 0.$$

Wiener was struck by this equation. In [177, p. 143] he writes:

The equations for radiation equilibrium in the stars belong to a type now known by Eberhard Hopf's name and mine. They are closely related to other equations which arise when two different physical regimes are joined across a sharp edge or a boundary, as for example in the atomic bomb, which is essentially the model of a star in which the surface of the bomb marks the change between an inner regime and an outer regime; . . .

From my point of view, the most striking use of Hopf-Wiener equations is to be found where the boundary between the two regimes is in time and not in space. One regime represents the state of the world up to a given time and the other regime the state after that time. This is the precisely appropriate tool for certain aspects of the theory of prediction, in which a knowledge of the past is used to determine the future.

As we shall see, this fascination with the H.W. equation had the interesting effect of leading Wiener to a treatment of prediction, which was at once more limited theoretically and more fruitful practically than that found independently by Kolmogorov (40) in 1941. In Kolmogorov's work the H.W. equation does not occur.

III. Nonlinear prediction

10. **On predictability in semi-exact sciences.** In fields such as meteorology the enormity of free coordinates, our ignorance of important factors and the sparseness of our data preclude us from establishing and solving strict dynamical equations. Even so we are able to make predictions. How are we to account for predicability in such "semi-exact" fields? Wiener's position on this important question and on the problem of effecting such prediction is stated in the essay [170]. This appeared in 1954 after much of his specialized work in prediction had been done. But on account of its comprehensive and thought-provoking nature we shall comment on its now, and then turn to specialized problems in prediction.[18]

Let Ω be the set of signals pertaining to the field in question, e.g., each $\omega \in \Omega$ may be the record of temperatures at different times (discrete or continuous) at a place P_ω. Assume that time is continuous, so that each ω is a function on $(-\infty, \infty)$, and that $t=0$ represents the present moment. Wiener held that while we cannot in general forecast the future value, $\omega(5)$ say, of a particular signal ω on the basis of our knowledge of the past, we can on this basis forecast the proportion p of the signals ω in Ω for which $\omega(5) \in S$, say, S being a given (Borel) set of numbers. To allow for the possibility that Ω is infinite, p has to be interpreted as a probability. Thus the first premiss of Wiener's theory is that *there exists a probability measure P on a Borel algebra $\mathfrak{B}$ over the signal-space Ω.* Another premiss is that the signals are "not tied down to any specific origin in time" [TS, p. 15], i.e., that Ω *is closed under translations* T_h: $\{T_h(\omega)\}(t) = \omega(t+h)$, $t, h \in (-\infty, \infty)$.

Now let p_t be the tth coordinate functional on Ω, i.e. $p_t(\omega) = \omega(t)$, and let $\mathfrak{B}_t$ be the Borel subalgebra of $\mathfrak{B}$ generated by the functions p_s, $s \leq t$. Consider a set $A \in \mathfrak{B}_t$, $t > 0$, such that $A \notin \mathfrak{B}_0$; e.g.

(10.1) $A = \{\omega: \omega(5) \in [a, b] \ \& \ \omega(7) \in [c, d]\} \in \mathfrak{B}_7.$

Notice that since $T_h^{-1}(\mathfrak{B}_t) = \mathfrak{B}_{t+h}$, we have

---

[18] [170] is couched in cryptic language. Some of the passages have to be read with considerable empathy. For instance, Wiener rightly emphasizes the central role of ergodicity in his theory (p. 249); but on p. 248, lines 6–8 and p. 250, 3rd paragraph, he gives the misleading impression that prediction can be carried out even when the set of moments accessible to observation is devoid of group structure.

$$T_7(A) = \{\omega: \omega(-2) \in [a, b] \ \& \ \omega(0) \in [c, d]\} \in \mathfrak{B}_0.$$

Without further hypotheses $P(A)$ cannot be found from our observations as these extend only up to $t=0$. But assuming that $P(B)$ can be found from observations for $B \in \mathfrak{B}_0$ and that the translations T_h preserve P measure, we can find $P\{T_7(A)\}$ from our observations and hence also $P(A) = P\{T_7(A)\}$.

Wiener invoked ergodicity to explain how $P(B)$, $B \in \mathfrak{B}_0$, can be found from observations in the past. Assuming that the translations T_h are ergodic, Birkhoff's theorem assures us that for almost all signals ω

$$\lim_{\tau \to \infty} \frac{1}{\tau} \int_{-\tau}^0 \chi_B(T_t\omega)dt = \int_\Omega \chi_B(\omega)P(d\omega) = P(B).$$

If $B = T_7(A)$ where A is as in (10.1), then $\int_{-\tau}^0 \chi_B(T_t\omega)dt$ is the Lebesgue measure of the set

$$\{t: -\tau \leq t \leq 0 \ \& \ \omega(t-2) \in [a, b] \ \& \ \omega(t) \in [c, d]\}.$$

This measure can, in principle, be found approximately from our record of the signal values $\omega(t)$ for $t \leq 0$. Thus a third premiss of Wiener's theory is that *the translations T_h, $-\infty < h < \infty$, form an ergodic, measure-preserving flow on $(\Omega, \mathfrak{B}, P)$.*

In short, Wiener's prediction theory is based on three postulates: (i) the signal space Ω is closed under all translations T_h, (ii) there exists a probability measure P on a Borel algebra $\mathfrak{B}$ over Ω, (iii) the translations T_h, $-\infty < h < \infty$, form a P-measure preserving and ergodic flow.

11. Nonlinear prediction (elementary standpoint). The next problem is to show how prediction is to be carried out on the basis of the postulates (i)–(iii) of §10. Now a probability space $(\Omega, \mathfrak{B}, P)$ in which Ω consists of signals, i.e. functions on $(-\infty, \infty)$, and satisfies postulate (iii) of §10 is simply a stationary, ergodic SP in its "coordinate representation," cf. Doob (20, I §6, X §1). As Doob has emphasized such a representation is dispensable. Viewed from a coordinate-free standpoint, *our problem is to carry out the best (nonlinear) prediction of a strictly stationary, ergodic SP.*

Wiener discussed this problem for real SP with discrete time in the joint paper [196] with the writer published in the Harald Cramer Volume. Adopting the RMS error criterion, which is standard in communication theory and in physics, cf. [TS, p. 13], it is shown that for a real-valued strictly stationary SP $(f_n)_{-\infty}^\infty$ over $(\Omega, \mathfrak{B}, P)$ such that $E(f_n) = 0$ the best prediction of f_ν with lead ν is the conditional expectation $E(f_\nu | \mathfrak{B}_0)$, where $\mathfrak{B}_0$ is the Borel subalgebra of $\mathfrak{B}$ spanned by the f_k for $k \leq 0$. Assuming that f_0 and therefore each f_k is in $L_\infty(\Omega, \mathcal{B}, P)$, we show that $E(f_\nu | \mathcal{B}_0)$ is the orthogonal projection of f_ν on the L_2-closure of the linear algebra $\mathcal{C}_0$ spanned by the f_k, $k \leq 0$, the so-called "nonlinear past of f_0." Finally, with the additional assumption that for distinct integers $n_1, \cdots, n_q$ the spectra of the distribution functions of the q-variates $(f_{n_1}, \cdots, f_{n_q})$ have positive q-dimensional Lebesgue measure, it is proved that

$$E(f_\nu | \mathcal{B}_0) = \underset{n \to \infty}{\text{l.i.m.}} \; Q_n \{f_0, f_{-1}, \cdots, f_{-m_n}\}$$

where m_n is a non-negative integer depending on n, and the Q_n are real polynomials in $m_n + 1$ variables, the coefficients of which are computable expressions of the moments of the SP. These moments can in principle be determined from time series data in the past on the basis of ergodicity.

Thus, under rather natural restrictions the nonlinear prediction problem is solvable, and indeed reducible to a linear problem, viz. the determination of the orthogonal projection of f_ν on the "nonlinear past," i.e. on a well-defined subspace of the Hilbert space $L_2(\Omega, \mathcal{B}, P)$. Of course, the practical difficulties of carrying out this solution are enormous, cf. §12.

12. **Nonlinear prediction (advanced standpoint).** The solution just outlined has the shortcoming of requiring the inversion of larger and larger matrices to get the different Q_n. Wiener felt that a better approach to nonlinear prediction required a deeper analysis of strictly stationary processes. In the linear case the corresponding analysis is that due to Wold and Kolmogorov. The problem now was to carry out a similar, but nonlinear, time-domain analysis of strictly stationary SP, and to follow this up with a characteristic functional analysis.

Wiener seems to have begun efforts in this direction around 1953, being guided to some extent by his earlier work on homogeneous chaos [108]. In 1956 he worked on the problem with G. Kallianpur at Calcutta. Their results appear in a technical report[19] and also in extremely diffuse form, and unfortunately without reference to Kallianpur, in the chapters entitled Coding and Decoding in Wiener's book [NPRT] published in 1958.[20] We must now examine this compelling theory, although it contains a lacuna as we shall see.

---

[19] Nonlinear prediction (with G. Kallianpur), Technical Report No. 1 (1956), Office of Naval Research, Cu-2-56-Nonr-266, (39)-CIRMIP, Project NR-047-015.

[20] The material also appears in a mimeographed write-up by E. J. Akutowicz of lectures given by Wiener at Massachusetts Institute of Technology after his return from India. This version differs from the Kallianpur-Wiener report (see footnote 19) in proof techniques but not in essential conception and results.

92 P. MASANI

Wiener and Kallianpur consider a real-valued strictly stationary SP $(f_k)_{-\infty}^{\infty}$ over $(\Omega, \mathcal{B}, P)$. For any set J of integers let $\mathcal{B}_J$ be the Borel algebra spanned by f_j, $j \in J$ and $\mathcal{F}_J$ be the family of Borel sets in the (not necessarily finite-dimensional) Euclidean space $\mathcal{R}^J$, $\mathcal{R}$ being the real number field.[21] Introduce the abbreviations

$$\mathcal{B}_n = \mathcal{B}_J, \quad \mathcal{F}_n = \mathcal{F}_J, \quad \text{where} \quad J = \{n, n-1, \cdots\}.$$

The (microscopic) conditional probability $P(f_0^{-1}(-\infty, \lambda] \,|\, \mathcal{B}_{-1})$ is a $\mathcal{B}_{-1}$-measurable function on Ω. Hence (cf. e.g. Dynkhin $(21, 1.5)$) there exists a $\mathcal{F}_0$-measurable function G on $\mathcal{R}^{\{0,-1,-2,\cdots\}}$ such that for each real λ, $G(\lambda, \cdot)$ is $\mathcal{F}_{-1}$-measurable and

$$P(\overset{-1}{f_0}(-\infty, \lambda] \,|\, \mathcal{B}_{-1})(\omega) = G(\lambda, f_{-1}(\omega), f_{-2}(\omega), \cdots).$$

Wiener's first hypothesis is that for almost all $\omega \in \Omega$,

(A) $G(\cdot, f_{-1}(\omega), f_{-2}(\omega), \cdots)$ *is strictly increasing and continuous on* $\mathcal{R}$.

This condition entails, of course, that $\mathcal{B}_{-1} \neq \mathcal{B}_0$, i.e. that the SP $(f_n)_{-\infty}^{\infty}$ is *nondeterministic*. Let

(12.1) $$g_0(\omega) = G(f_0(\omega), f_{-1}(\omega), f_{-2}(\omega), \cdots).$$

Obviously the range of g_0 is essentially contained in $[0, 1]$. It can be shown that g_0 is independent of $f_{-1}, f_{-2}, \cdots$ and is uniformly distributed. Letting Φ be the $(0, 1)$ normal distribution function on $\mathcal{R}$ and $h_0 = \Phi^{-1}(g_0)$ it follows that h_0 is $(0, 1)$ normally distributed and independent of $f_{-1}, f_{-2}, \cdots$, and

$$h_0(\omega) = H(f_0(\omega), f_{-1}(\omega), \cdots), \qquad H = \Phi^{-1} \circ G.$$

Defining

(12.2) $$h_n(\omega) = H(f_n(\omega), f_{n-1}(\omega), \cdots), \qquad -\infty < n < \infty,$$

it follows that $(h_n)_{-\infty}^{\infty}$ is an independent, $(0, 1)$ normally distributed SP, and h_n is $\mathcal{B}_n$-measurable and independent of $f_{n-1}, f_{n-2}, \cdots$. We can also show that $\mathcal{B}_n = \mathcal{B}(h_n, f_k, k < n)$. Since

(i) h_n *is* $(0, 1)$ *normally distributed*,

(12.3) (ii) h_n *is* $\mathcal{B}_n$*-measurable and independent of* $f_{n-1}, f_{n-2}, \cdots$,

(iii) $\mathcal{B}_n = \mathcal{B}(h_n, f_{n-1}, f_{n-2}, \cdots)$,

we are entitled to regard h_n as the "nth innovation" of the SP $(f_n)_{-\infty}^{\infty}$. Wiener and Kallianpur thus established the existence of (*non-*

---

[21] $\mathcal{F}_J$ is the Borel algebra spanned by all sets of the form $X_{j \in J} B_j$, where B_j are Borel subsets of $\mathcal{R}$.

linear) *innovations* for a stationary SP obeying the condition (A) of strong nondeterminism.[22]

Wiener's next objective was to seek hypotheses which would ensure that his original SP $(f_n)_{-\infty}^{\infty}$ be a nonlinear, one-sided moving-average of his innovation SP:

$$(12.4) \qquad f_n = F(h_n, h_{n-1}, \cdots), \qquad -\infty < n < \infty.$$

An easy generalization of 12.3(iii) is

$$(12.5) \quad \mathcal{B}_n = \mathcal{B}(h_n, h_{n-1}, \cdots, h_{n-k-1}, f_{n-k-2}, \cdots), \qquad n, k \geqq 1.$$

Wiener and Kallianpur now impose their second hypothesis, viz. the SP $(f_n)_{-\infty}^{\infty}$ is *purely nondeterministic*, i.e.

$$(B) \qquad\qquad \mathcal{B}_{-\infty} = \bigcap_{n \geqq 0} \mathcal{B}_{-n} = \{\text{void set}, \Omega\}.$$

From (12.5) and (B) they conclude (letting $k \to \infty$) that $\mathcal{B}_n = \mathcal{B}(h_k, k \leqq n)$, and thence (12.4).

In 1959, about a year after the publication of [**NPRT**], M. Rosenblatt (**63**) gave an example in which the hypotheses (A) and (B) are fulfilled but (12.4) fails. The inference of $\mathcal{B}_n = \mathcal{B}(h_k, k \leqq n)$ from (12.5) and (B) is thus incorrect. But once (12.4) is properly established (say by suitably strengthening (A) and (B), or by changing the definition of g_0) we would have a theory useful for prediction. Thus with $f_k \in L_2(\Omega, \mathcal{B}, P)$ and the RMS criterion we would get

$$E(f_n \mid \mathcal{B}_0) = H(\underbrace{0, \cdots 0}_{n \text{ times}}, h_0, h_{-1}, \cdots), \qquad n \geqq 0.$$

Once H is known our predictor would be determined. Indeed we could get from (12.4) any statistical parameter of the conditional distribution of f_n relative to $\mathcal{B}_0$, e.g. the median.

13. Unsolved questions. Wiener has bequeathed to posterity the important problem of strengthening his hypotheses (A), (B), (§12) or of changing the definition (12.1) of g_0 so as to ensure both (12.2) and (12.4). Also left to us is the extension of this theory to the continuous parameter case. Here the absence of an atomic time-unit makes the problem of defining nonlinear innovations extremely hard; obviously all we may expect are virtual or differential innovations.

Also remaining on the agenda is the implementation of Wiener's idea expressed on the last page of his brilliant essay [**170**] of affecting

---

[22] Wiener and Kallianpur seem to have been unaware that their technique for defining g_0 is an interesting "infinite" adaptation of one used by P. Levi (**43**, Ch. VI) in deriving from a sequence $(f_n)_1^{\infty}$ a sequence of independent uniformly distributed varieties g_n such that $f_n = F_n(g_1, \cdots, g_n)$.

a nonlinear prediction of a (simple) SP by carrying out a linear prediction of a suitably derived infinite dimensional SP. There also remains the job of supplementing nonlinear, time-domain analysis by a characteristic functional analysis.

IV. LINEAR PREDICTION THEORY

14. The linear prediction and filtering problems. When the strictly stationary SP $(f_k, -\infty < k < \infty)$, where k can be integral or real, is Gaussian with zero expectations and $f_k \in L_2(\Omega, \mathcal{B}, P) = \mathcal{H}$, the best (nonlinear) predictor $E(f_\nu | \mathcal{B}_0)$, $\nu > 0$, considered in §11, turns out to be the orthogonal projection $\hat{f}_\nu$ of f_ν on the (closed, linear) subspace $\mathfrak{M}_0$ of $\mathcal{H}$ spanned by f_k, $k \leq 0$. When $(f_k)_{-\infty}^{\infty}$ is not Gaussian but merely stationary, $\hat{f}_\nu$ provides the first, i.e. linear, approximation to the best prediction of f_ν. On both counts the problem of finding the orthogonal projection of f_ν on $\mathfrak{M}_0$ is important. This is the *linear prediction problem* with lead ν, independently conceived and studied by Wiener and Kolmogorov.

When the time-signals of our stationary SP $(f_k, -\infty < k < \infty)$ are contaminated messages we may at the linear level assume that $f_k = f_k^1 + f_k^2$, where the variates f_k^1, f_k^2 represent the (*pure*) *message* and (*pure*) *noise*, respectively. Our problem is to find the best linear approximation to the message-variate f_ν^1 in terms of the signal variates f_k, $k \leq 0$, i.e. the orthogonal projection $\hat{f}_\nu^1$ of f_ν^1 on the subspace $\mathfrak{M}_0$ spanned by f_k, $k \leq 0$. This is the *linear filtering problem* (with lead or lag ν), also originally conceived by Wiener.

How Wiener was drawn to these problems by his involvement with antiaircraft fire control and noise filtration in radar during the war, and how his mathematical background fitted him ideally for this task are recounted in his [204], [177], [TS]. His work in this field falls rather naturally into two periods. The first, 1940–1943, culminated in his [TS], which was completed in early 1942 but appeared in declassified form only in 1949. The second period, 1949–1959, began with his addresses [143], [156] to the CNRS in Paris, and the International Congress of Mathematicians at Harvard, and ended with his papers on multivariate prediction.

During the first period Wiener was interested chiefly in getting autoregressive integral or series representations for the predictor and filter. He used variational techniques and solved the resulting Hopf-Wiener type integral equations. He accomplished all this without the use of too much theory by operating at what Yaglom [71, p. vii] has aptly called "a heuristic level of rigor." Some of the problems he tackled were formidable, but his work in this period lacked the theoretical strength and completeness of that of Kolmogorov (40).

On the other hand during the second period Wiener adopted a more theoretical approach. He undertook time-domain and spectral analysis, leaning more and more on abstract theory, especially Hilbert spaces.

15. **Linear prediction and filtering (first period).** Despite its hard mathematical content, which earned it the nickname "the yellow peril," [TS] has had a wide influence in engineering circles. This stemmed from its wealth of engineering insights and its emphasis on the quick solution of engineering problems without much fuss over rigor. The urgencies of war as well as Wiener's long-standing fascination with such problems and the functional equations to which they led had much to do with this attitude. As there is still some doubt as to the locus standi of the prediction and filtering techniques used in [TS] we shall briefly review them here in the light of the current theory of stochastic processes.

Consider a stationary, purely nondeterministic SP $(f_t, t$ real), $f_t \in L_2(\Omega, \mathcal{B}, P)$. As indicated in §10, we can estimate the covariances $\phi(t) = (f_t, f_0)$ from time-signal data in the past on the hypothesis of ergodicity, which Wiener freely made. Thus the covariance function ϕ is "known." Our problem is to find for a given $h > 0$, the orthogonal projection $\hat{f}_h$ of f_h on $\mathfrak{M}_0$, the closed subspace spanned by f_t, $t \leq 0$. Wiener studied this problem in the mathematically restrictive but practically important case in which $\hat{f}_h$ is given by an autoregressive Stieltjes integral:

$$(15.1) \qquad f_h = \int_0^\infty f_{-\tau} dw(\tau)$$

where w is of bounded variation on $[0, \infty)$. (As yet, no nice necessary and sufficient criterion for the validity of such a representation of $\hat{f}_h$ is known.) Since $f_h - \hat{f}_h \perp f_{-t}$, $t \geq 0$, we readily get on taking inner products

$$(15.2) \qquad \phi(t + h) = \int_0^\infty \phi(t - \tau) dw(\tau), \qquad t \geq 0.$$

This is a *Hopf-Wiener Stieltjes integral equation of the first kind*, by solving which the "unknown" function w is to be found. Wiener arrived at this equation [TS, (2.021)] by a longer variational approach.

To solve (15.2) let us proceed heuristically. We extend w to $(-\infty, \infty)$ by defining it to be zero on $(-\infty, 0)$ and define

$$\Psi_h(t) = \phi(t + h)\chi_{[0,\infty)}(t), \qquad -\infty < t < \infty,$$

$$(15.3)$$

$$u(t) = \int_{-\infty}^\infty \phi(t - \tau) dw(\tau) - \Psi_h(t), \qquad -\infty < t < \infty.$$

Denoting as usual the indirect Fourier transform by a tilde, noting that $\tilde{\phi}(\lambda) = F'(\lambda)$, where F' is the spectral density of the SP, and letting

$$(15.3') \qquad W(\lambda) = \int_0^\infty e^{i\lambda t} dw(t),$$

we get from (15.3)

$$(15.4) \qquad \tilde{u}(\lambda) = F'(\lambda)W(\lambda) - \tilde{\Psi}_h(\lambda), \qquad -\infty < \lambda < \infty.$$

Now Wiener knew that his hypothesis that the SP is *nondeterministic*, i.e. $f_h \notin \mathfrak{M}_0$, entails the condition

$$(15.5) \qquad \{\log F'(\lambda)\}/(1 + \lambda^2) \in L_1(-\infty, \infty).$$

(Actually the equivalence of (15.5) to nondeterminism was proved only later by Karhunen (35).) From (15.5) and the Paley-Wiener Theorem [92, pp. 16–17]

$$(15.6) \qquad F'(\lambda) = |\Phi(\lambda)|^2 \quad \text{a.e.} \quad \text{on } (-\infty, \infty),$$

where Φ has a holomorphic extension to the upper half plane Δ_+, this extension being in the Hardy class H_2. One can choose a Φ which is free from zeros in Δ_+. Then $1/\Phi$ will be holomorphic on Δ_+ and $1/\bar{\Phi}$ on the lower half plane Δ_-. Also since $u=0$ on $[0, \infty)$ and $w=0$ on $(-\infty, 0]$, we see that $\tilde{u}$ and W have holomorphic extensions to Δ_-, Δ_+ respectively. Hence on dividing by $\bar{\Phi}$ in (15.4) we find that

$$\Phi(\lambda)W(\lambda) = [\tilde{\Psi}_h(\lambda)/\bar{\Phi}(\lambda)]_+$$

where $[\ ]_+$ denotes the operation of cutting off the negative frequencies.[23] It easily follows that

$$(15.7) \qquad W(\lambda) = \frac{1}{\Phi(\lambda)}[\tilde{\Psi}_h(\lambda)/\bar{\Phi}(\lambda)]_+ = \frac{1}{\Phi(\lambda)}[e^{-ih\lambda}\Phi(\lambda)]_+.$$

This yields W, from which w or a suitably normalized version of w can be retrieved by inversion:

$$w(t) - w(0) = \lim_{A \to \infty} \frac{1}{2\pi} \int_{-A}^A \frac{1 - e^{-it\lambda}}{i\lambda} W(\lambda) d\lambda.$$

The exact hypotheses needed to validate this heuristic solution have not been discovered. Wiener showed that it goes through in case the spectral density F' is a rational function P/Q with $\deg P \leq \deg Q$. This case, though mathematically restrictive, is important in applications.

---

[23] I.e. $[\int_{-\infty}^\infty e^{it\lambda}g(t)dt]_+ = \int_0^\infty e^{it\lambda}g(t)dt.$

The above-mentioned treatment of the prediction problem was in marked contrast to that of Kolmogorov which covered the entire field of prediction of discrete parameter univariate processes, first in the time-domain with heavy emphasis on the fundamental one-sided moving average representation discovered by Wold (66) in 1938, and then in the spectral domain by exploiting powerful theorems on Hardy class functions. Kolmogorov showed that there is an isomorphism between the time and spectral domains. It may be shown that if we define W not by (15.3′) but by (15.7), taking the Φ therein to be the *optimal* (or outer) function satisfying (15.6), then $\hat{f}_h$ and W are isomorphs, cf. e.g. [186, 4.11]. Yaglom, Darlington and other workers interested in the engineering side have developed algorithms for computing the predictor in the frequency domain starting directly from this fact and without assuming the autoregressive representation (15.1) in the time-domain. The Hopf-Wiener equation (15.2) is thus avoided (68), (16).

Wiener approached linear filtering in much the same spirit as linear prediction. Let $f_t = f_t^1 + f_t^2$, where f_t^1, f_t^2 represent the pure message and pure noise, respectively. As before we assume as known the auto- and cross-covariance functions ϕ, ϕ_1:

$$\phi(t) = (f_t, f_0), \qquad \phi_1(t) = (f_t^1, f_0).$$

As in (15.1) Wiener assumed that for any real h, the projection $\hat{f}_h^1$ of f_h^1 on the subspace $\mathfrak{M}_0$ spanned by f_t, $t \leq 0$ is given autoregressively:

$$(15.1') \qquad \hat{f}_h^1 = \int_0^\infty f_{-\tau} dw(\tau).$$

Instead of (15.2) he now gets the Hopf-Wiener type equation

$$(15.2') \qquad \phi_1(t + h) = \int_0^\infty \phi(t - \tau) dw(\tau), \qquad t \geq 0,$$

cf. [TS, (3.20), (3.52)]. Keeping to our former notation except for using ϕ_1 instead of ϕ in defining $\Psi_h(t)$, we get as before

$$(15.7') \quad W(\lambda) = \frac{1}{\Phi(\lambda)} \, [\tilde{\Psi}_h(\lambda)/\overline{\Phi}(\lambda)]_+ = \frac{1}{\Phi(\lambda)} \, [e^{-ih\lambda} F_1(\lambda)/\overline{\Phi}(\lambda)]_+,$$

where F_1 is the cross-spectral distribution of the f_t^1- and f_t-processes. From (15.7′) w can be retrieved under suitable assumptions. Wiener's approach was again rather pragmatic. As in prediction one may show without recourse to (15.1′) and (15.2′) that the function W in (15.7′) is the spectral isomorph of the filtration $\hat{f}_h^1$.

16. Linear prediction and filtering (second period). As mentioned in §14, around 1949 Wiener began to veer towards a more theoretical approach to weakly stationary SP's. In his addresses [143, 156] he dealt with moving averages in the time-domain (unfortunately without reference to Wold) and followed this by spectral analysis. This trend persists in the papers [172, 176] in which Hilbert space techniques are increasingly used. But Wiener's analysis was incomplete. For instance, the remote past and nonabsolutely continuous spectra were not considered. Indeed, so thorough had been Kolmogorov's treatment of univariate prediction in the discrete case (40) that there was little left to do.

In [156, 176] Wiener also discussed the prediction of continuous parameter, weakly stationary processes. This aspect of his work is best broached by a quotation [176, p. 183]: "Let the unitary transformation T^{-1} carry $f(\alpha)$ into $g(\alpha)$ if and only if for all n

$$\int_{-\infty}^{\infty} f(S^{-t}\alpha)l_n(t)dt = \int_{-\infty}^{\infty} g(S^{-t}\alpha)l_{n-1}(t)dt."$$

In this interesting but rather cryptic sentence the l_n are Laguerre functions and $(S^t,\ t$ real) is a measure-preserving flow on $([0,\ 1]$, Borel, Leb.). When deciphered the sentence reads: "Let T be the Cayley transform of H, where iH is the infinitesimal generator of the unitary group $(S_t,\ t$ real) induced by the S^t-flow: $S_t(f)=f \circ S^t$." It seems that Wiener did not realise this; even so he took the important step of associating with his continuous parameter SP $(S_t(f),\ t$ real) the discrete parameter SP $(T^n(f),\ n=$integer), and so bringing the discrete theory to bear on the continuous. But he did not pursue this fine idea systematically, and his work on the continuous case during this period is on the whole rather sketchy.

In [143, 176] Wiener also considered linear filtering with discrete time. He employed moving averages and spectral theory to prove results obtained either heuristically or under restrictions in earlier work.

17. Multivariate prediction. No sooner had Wiener tackled the prediction and filtering problems for univariate processes, he turned to the corresponding problems for q-variate processes $(f_t,\ t$ real), where each f_t is a q-ple vector $(f_t^j)_{j=1}^q$ with $f_t^j \in L_2(\Omega,\ \mathcal{B},\ P)$, and the Gram matrix $(f_s,\ f_t) = [(f_s^i,\ f_t^j)] = \phi(s-t)$ depends only on the difference $s-t$. In [TS, Ch. IV] he assumed, in keeping with his approach for the case $q=1$, that the predictor $\hat{f}_h$ has an autoregressive Stieltjes integral representation. He then got instead of the single Hopf-Wiener equation (15.2) a system of q such equations in q unknowns

[TS, (4.105)]. This system yields on solution the first component of $\hat{f}_h$. To get $\hat{f}_h$ itself we must solve a system of q^2 such equations in q^2 unknowns, i.e. a single matrix Hopf-Wiener equation

$$\phi(t + h) = \int_{-\infty}^{\infty} \phi(t - \tau)dw(\tau)$$

where w is the (unknown) $q \times q$ matricial weighting. Wiener's treatment of the solution of this equation was again largely heuristic. It can be carried through only under rather severe restrictions on the process, which, however, are fulfilled in many cases of practical interest.

A more theoretical approach to multivariate prediction occurs in Wiener's address [156], and is amplified in the paper [172] dedicated to Plancherel. Wiener considered a discrete parameter, purely non-deterministic, bivariate SP of full rank, i.e. one for which $q = 2$, the so-called remote past $M_{-\infty}$ is $\{0\}$ and the 1-step prediction error matrix has rank 2. By alternating projections in Hilbert space Wiener obtained expressions for the innovation vectors. But questions of existence and those of computation were not kept apart, and this delimited the work. Thus Wiener was able to show that such a process has a spectral density F' and that F' is factorizable:

$$(17.1) \qquad F' = \Phi\Phi^*, \quad \text{a.e.,} \quad \Phi \sim \sum_{k=0}^{\infty} C_k e^{ki\theta} \in L_2(C),$$

where C is the unit circle $\{z: |z| = 1\}$. But he was not quite able to show that the condition

$$(17.2) \qquad \log \det F' \in L_1(C)$$

suffices for such factorization.[24] In [172] Wiener also discussed a similar factorization for unitary matrix-valued functions, but his proof is incorrect and the result itself is in doubt.

The work done up to this point by Wiener and others had not cleared up the basic questions of q-variate prediction theory, and clearly pointed to the need for a systematic study of these questions unencumbered by algorithmic considerations. Wiener began this work in collaboration with the writer in 1955–1956 at Calcutta. This research [185, 186], deals with the general case first and only later with special cases.[25] As the subject is rather technical we shall assume in what follows that the reader is familiar with the basic concepts.

---

[24] His argument fails when the angle between the past subspaces of the component processes is zero.

[25] [185] owes much to Zasuhin's brilliant Doklady note (72), which announced a chain of general results on q-variate processes.

[185] begins with the time-domain analysis of q-variate processes of rank ρ, $1 \leq \rho \leq q$. Following Zasuhin ρ is defined to be the rank of the 1-step prediction error Gram matrix

$$G = (\boldsymbol{g}_0, \boldsymbol{g}_0) = [(g_0^i, g_0^j)]$$

where $\boldsymbol{g}_0 = (g_0^i)_{i-1}^q$ is the 0th innovation vector. The Wold or rather *Wold-Zasuhin decomposition* is established, and necessary and sufficient criteria given for pure nondeterminism, i.e. for $\boldsymbol{M}_{-\infty} = \{0\}$. The paper also contains spectral analysis. Perhaps the most fundamental result is the *determinantal extension of the Szegö-Kolmogorov identity*:

$$(17.3) \qquad \log \det \boldsymbol{G} = \frac{1}{2\pi} \int_0^{2\pi} \log \det \boldsymbol{F}'(e^{i\theta}) d\theta,$$

where $\boldsymbol{F}$ is the $q \times q$ matricial spectral distribution of the SP. This result, first stated by Whittle (65), is proved by exploiting theorems on the Hardy classes as well as the concavity of the functional log det on the space of $q \times q$ non-negative, hermitian matrices. (We get a generalized Jensen inequality.) From (17.3) it follows at once that (17.2) is the n.a.s.c. that $\rho = q$, i.e. that the SP be of *full rank*. We show next that when $\rho = q$ the absolutely continuous and nonabsolutely continuous parts $\boldsymbol{F}_a$, $\boldsymbol{F}_b$ of $\boldsymbol{F}$ are the spectral distributions of the purely nondeterministic part $(\boldsymbol{u}_n)_{-\infty}^\infty$ and the deterministic part $(\boldsymbol{v}_n)_{-\infty}^\infty$ of the Wold-Zasuhin decomposition of $(\boldsymbol{f}_n)_{-\infty}^\infty$—the so-called *concordance of Wold-Zasuhin and Lebesgue-Cramer decompositions*. By appeal to Cramer's criterion that a matrix-valued function be the spectral distribution of a SP (12), we also show that if $\boldsymbol{F}' \geq 0$, $\boldsymbol{F}' \in L_1(C)$ and satisfies (17.2), then $\boldsymbol{F}'$ admits a factorization of the form (17.1), where moreover the factor Φ is *optimal* ("outer" in Beurling's terminology), i.e.,

$$(17.4) \qquad \boldsymbol{C}_0 \geqq 0, \qquad \log \det \boldsymbol{C}_0 = \frac{1}{2\pi} \int_0^{2\pi} \log \det \boldsymbol{F}'(e^{i\theta}) d\theta.^{26}$$

In [186] an isomorphism is established between the time and spectral domains of a q-variate purely nondeterministic SP of full rank q. But the primary goal is the derivation of an algorithm for the linear prediction $\hat{\boldsymbol{f}}_\nu$ of $\boldsymbol{f}_\nu$ with lag ν. $\hat{\boldsymbol{f}}_\nu$ turns out to be the isomorph of the function $\boldsymbol{Y}_\nu$:

$$(17.5) \qquad \boldsymbol{Y}_\nu(e^{i\theta}) = [e^{-\nu i\theta}\Phi(e^{i\theta})]_{0+}\Phi^{-1}(e^{i\theta})$$

---

26 Some of these results were obtained independently by M. Rosenblatt, Helson and Lowdenslager, and Iu. A. Rosanov.

where Φ is the unique function satisfying (17.1) and (17.4). The problem is therefore to find an algorithm for the computation of this "optimal factor" of F'. In case $q=1$

$$\Phi(z) = \exp\left\{\frac{1}{2\pi}\int_0^{2\pi}\frac{e^{i\theta}-z}{e^{i\theta}+z}\log F'(e^{i\theta})d\theta\right\}, \qquad |z|<1.$$

But for $q>1$ no such closed-form expression involving the log is available since matrix multiplication is noncommutative, and all one can hope for is some iterative procedure for finding Φ. Our algorithm emerges when Wiener's approach for $q=2$ based on alternating projections [172] is judiciously fused with a technique of noncommutative factorization used earlier by the writer (46). Under the assumptions $\lambda I \leq F'(e^{i\theta}) \leq \lambda' I, \ 0<\lambda\leq\lambda'<\infty$, we show that

$$(17.6) \quad \sqrt{G}\Phi^{-1} = I - M_+ + (M_+M)_+ - \{(M_+M)_+M\}_+ + \cdots,$$

where

$$M = \frac{2}{\lambda+\lambda'}F' - I$$

and the subscript $+$ denotes the operation of cutting off all but positive frequencies. For $\hat{f}_{\nu}$ itself we obtain a mean-convergent auto-regressive series

$$(17.7) \qquad \hat{f}_{\nu} = \sum_{k=0}^{\infty} E_{\nu k}f_{-k},$$

where the $E_{\nu k}$ are finite sums of the Fourier coefficients of Φ^{-1} and Φ.

A natural sequel to the foregoing study is that of purely nondeterministic processes of degenerate rank: $1\leq\rho<q$. In this case F is absolutely continuous,

$$(17.8) \ \det F' = 0, \text{ a.e.}, \quad F' = \Phi\Phi^*, \text{ a.e.}, \quad \Phi \sim \sum_{k=0}^{\infty} C_k e^{ki\theta} \in L_2(C).$$

What conditions must F' satisfy in order that such a *"degenerate rank" factorization* be possible? In [197] Wiener and the writer gave a complete answer for the bivariate case, $q=2$: with $F=[F_{ij}]$,

$$(17.9) \quad \begin{aligned} &\log F'_{ii} \in L_1(C), \qquad i = 1 \text{ or } 2, \\ &F'_{ji}/F'_{ii} = \textit{radial limit of a beschränktartige function}, \quad j \neq i. \end{aligned}$$

At about the same time Wiener wrote a paper [194] with Akutowicz in which the "full rank" factorization theorem of [185] is reproved *ab initio*. The difficulty which Wiener had encountered in his **earlier** work [172, 176] is avoided by abandoning the method of alternating projections in favor of a technique suggested by the time-domain analysis of q-variate processes given in [185].

18. **The continuation of Wiener's work on prediction.** The work of Kolmogorov and Wiener on prediction has had widespread influence. Activity in the field has been especially vigorous from the time when Wiener's work on multivariate prediction and that of Rosanov (60) in Russian appeared. In the following brief account of these developments we shall only mention work closely related to Wiener's. Our order will be logical rather than chronological.

During 1947–1950 Karhunen (35) and Hanner (26) extended Kolmogorov's work to univariate continuous parameter processes, and thereby settled important questions which Wiener had bypassed in his early, heuristic attacks.[27] But the techniques used were somewhat ad hoc in nature. Now (cf. §16) Wiener [176] had suggested the possibility of associating with the weakly stationary, continuous parameter SP ($U_t f$, t real) the weakly stationary, discrete parameter SP ($V^n f$, $n =$ integer), where V is the Cayley transform of H, iH being the infinitesimal generator of the unitary group (U_t, t real). J. Robertson and the writer (55) showed that such association results in a coherent and simple development of the entire theory. In particular, the troublesome process of orthogonal increments (ξ_t, t real), the differentials of which are the (virtual) innovations of the f_t-process, is easily obtained.

On the engineering side, Bode and Shannon (7) simplified Wiener's early version of prediction and filtering [TS] using circuit theory concepts. The circuit theory point of view also led Darlington (16) to a simpler form of the theory adaptable to applications exemplifying rational spectra. Zadeh and Ragazinni (72) modified the Wiener theory to cover the case in which the known data is confined to a bounded time-interval in the past. Yaglom (68) put some of this work on a rigorous footing. J. Chover (9), (10) made an interesting mathematical analysis of the case in which such prediction in terms of a bounded interval is representable autoregressively by a Stieltjes integral, cf. (15.1). Dolph and Woodbury (18) studied the predictor and filter, again on the basis of a bounded time-interval, for SP's, the signals of which are generated by driving linear differential systems by white noise. Kalman and Bucy (34) also considered this problem, but from the general standpoint of stochastic control theory.

To turn to the q-variate theory, the isomorphism between the time

---

[27] The state of the subject at this point (1950) is well exposed in Doob's book (20, Ch. XII).

and spectral domains established in [186] under the hypothesis of full rank and pure nondeterminism was obtained without restriction by M. Rosenberg (59) and Rosanov (62'). Research in the field was thereby freed from reliance on unnecessary hypotheses. Matveev (56) studied degenerate rank factorizations of matrix-valued functions for any $q \geq 1$. His condition for the factorization (17.8) with rank $\Phi = \rho$, $1 \leq \rho < q$ is a direct generalization of the condition (17.9) proved in [197] for $q = 2$, $\rho = 1$. Somewhat less tractable conditions for degenerate rank factorization were found by Helson and Lowdenslager (28). The writer (47) showed that the concordance of the Wold-Zasuhin and Lebesgue-Cramer decompositions, established in [185] in the full rank case, breaks down when $1 \leq \rho < q$. For the case $q = 2$, $\rho = 1$, he gave necessary and sufficient conditions for the prevalence of concordance, which J. Robertson (58) generalized to $q \geq 1$, $1 \leq \rho < q$, by first proving an elegant result on the ranges of the matrices $F_x'(e^{i\theta}), F_y'(e^{i\theta}), F_z'(e^{i\theta})$, where the x_n-process is the sum of orthogonally related processes y_n and z_n. Some of Robertson's results were duplicated independently in China by Jang Ze-pei (31).

On the computational side, the writer (48) showed that the algorithm (17.6) for finding the optimal factor Φ of F' satisfying (17.1) and (17.4) extends to the case in which the reciprocal matrix $(F')^{-1}$ and the quotient $\lambda'(\cdot)/\lambda(\cdot)$ of the largest to the smallest eigenvalue of F' are in $L_1(C)$. He also showed that the autoregressive series (17.7) for the predictor is available under the weak hypotheses $F' \in L_\infty(C)$, $(F')^{-1} \in L_1(C)$. Yaglom (70) developed algorithms for prediction for continuous parameter q-variate processes with rational spectra.

Gangoli (24) developed a q-variate prediction theory valid for $q = \infty$. He defined a SP as a bisequence of bounded linear operators F_n from Hilbert spaces $\mathcal{K}$ to $\mathcal{K}'$, where dim $\mathcal{K} = q$, and showed that many of the definitions, result and proofs given in [185] carry over to the case $q = \infty$. Gangoli also considered the factorization $W' = \Phi\Phi^*$, where W' is a function on the unit circle C, the values of which are non-negative, hermitian operators on $\mathcal{K}$ to $\mathcal{K}$, and Φ has a one-sided Fourier development. This factorization had been treated earlier by Devinatz (17), who showed that a sufficient condition is that $\log \lambda(\cdot) \in L_1(C)$, where $\lambda(e^{i\theta})$ is the g.l.b. of the spectrum of $W'(e^{i\theta})$, and by Lax (42), who showed that the literal generalization of the Szegö condition, $\log W' \in L_1(C)$, is inadequate.

Among the more distant work influenced by the contributions of Wiener and Kolmogorov we should mention (i) the extension of prediction theory to stationary random distributions due to K. Ito (30), Rosanov (61), and Balagangadharan (1); (ii) Cramer's extension of prediction theory to nonstationary processes in both discrete and continuous time (**13, 14, 15**). (iii) the firm initiation of a theory of linear least squares interpolation due to Yaglom and Rosanov (**39, 67, 62**).

All in all, the pioneering work of Wiener and Kolmogorov in prediction has stimulated much valuable mathematical activity in many lands.

19. **The ramifications of prediction theory.** It has become increasingly clear that linear prediction theory is a part of the general theory of one-parameter semi-groups of isometries on a Hilbert space. For instance, the Wold Decomposition of a weakly stationary, discrete parameter SP $(U^n f)_{-\infty}^{\infty}$ follows at once from the general equation, cf. Halmos (**25**),

$$(19.1) \qquad \mathfrak{IC} = \bigcap_{n=1}^{\infty} V^n(\mathfrak{IC}) + \sum_{k=1}^{\infty} V^k(R^{\perp})$$

which holds for any Hilbert space $\mathfrak{IC}$ and any isometry V on $\mathfrak{IC}$ onto $R \subseteq \mathfrak{IC}$. (Just take $\mathfrak{IC}$ to be the present and past subspace $\mathfrak{M}_0$ of the given SP and V to be the restriction of U^* to $\mathfrak{M}_0$.) The general theory of isometric semi-groups also embraces parts of functional analysis such as the theory of shifts, the Hardy class functions, etc. For instance, Beurling's Theorem (**3**, IV) on shift-invariant subspaces of the Hardy class H_2 also follows from (19.1) when $\mathfrak{IC}$ and V are properly chosen.

Now many of the ideas and techniques suggested by prediction theory have proved useful in these related areas, and so the light emanating from Wiener's work on prediction has reached well beyond the confines of prediction theory proper. As examples of such distant ramifications of prediction theory we may mention (i) the deduction of the optimal-residual[28] factorization of a function f in the Hardy class H_2 from the Wold Decomposition of the SP $(e^{-ki\theta}f)_{k=-\infty}^{\infty}$ (**49**); (ii) Lax's vectorial extension of Beurling's Theorem on shift-invariant subspaces using techniques suggested by q-variate prediction theory, (**41**); (iii) the canonical factorization of matrix-valued functions in the Hardy class H_2 obtained by the writer, again employing the ideas of q-variate theory, (**49–53**); (iv) the writer's enunciation and proof of the analogue of (19.1) for a continuous parameter semi-group $(S_t, t \geq 0)$ of isometries, in which he was guided by the situation in continuous time processes, (**54**); (v) the Hardy class, theory on the torus due to Helson and Lowdenslager (**27**). Some of the recent work of Halmos and his collaborators, of Lowdenslager,

---

[28] Outer-inner, in Beurling's terminology.

Helson and others on the invariant subspace problem also falls into this category.

There are other regions of functional analysis where the viewpoint of prediction theory might fructify. An instance is the study of the operator identities of G. Baxter (2). Some of these identities were actually encountered in [186] in the derivation of the algorithm (17.6). Another instance is the theory of stationarity and prediction for individual time signals and certain associated algebras due to H. Furstenberg (22).

V. Filter theory

20. Linear filters. Realizability. The only linear filters which Wiener considered at length were time-invariant ones which possess a differential *weighting* K in the time domain, i.e. filters for which the response g to an input signal f is given by

$$(20.1) \quad g(t) = \int_{-\infty}^{\infty} dK(t - \tau)f(\tau) = (dK*f)(t), \qquad t \in (-\infty, \infty)$$

where K is a function of bounded variation on $(-\infty, \infty)$ and $*$ denotes convolution. An especially important subclass of such filters is that for which K is absolutely continuous on $(-\infty, \infty)$. Writing $K' = -W$, we then have

$$(20.2) \quad g(t) = \int_{-\infty}^{\infty} W(t - \tau)f(\tau)d\tau = (W*f)(t), \qquad t \in (-\infty, \infty)$$

where $W \in L_1(-\infty, \infty)$. We shall refer to (20.1) and (20.2) as the $dK*$ and $W*$ filters.

To "know" the filters is to know the weightings K or W. A major problem of linear filter theory is to determine these weightings from a comparative analysis of the (observable) input and output signals f and g. Wiener made several contributions to this problem, cf. §§22, 23 below.

All physically realizable filters are *causal* (or *retrospective* or *non-anticipative*) in the sense that for each t

$$f_1 = f_2 \text{ on } (-\infty, t) \implies g_1 = g_2 \text{ on } (-\infty, t).$$

Obviously the filters $dK*$, $W*$ will be causal if and only if $K = 0 = W$ on $(-\infty, 0)$. In this case (20.1), (20.2) reduce to

$$(20.1') \qquad g(t) = \int_{-\infty}^{t} dK(t - \tau)f(\tau), \qquad t \in (-\infty, \infty),$$

$$(20.2') \qquad g(t) = \int_{-\infty}^{t} W(t - \tau)f(\tau)d\tau, \qquad t \in (-\infty, \infty).$$

One of Wiener's first contributions was to give the *spectral necessary and sufficient condition for causality of the filter* W_*, viz.

$$(20.3) \qquad \int_{-\infty}^{\infty} \{ |\,|\log|\,\tilde{W}(\lambda)\,|\,|\,/(1+\lambda^2)\} \, d\lambda < \infty,$$

where $\tilde{W}$ is the indirect Fourier transform of W, cf. (1.1). This condition emerges from his joint work with Paley (§8), and engineers refer to it as the *Paley-Wiener condition*.

In case the filter W_* is causal, i.e. $W = 0$ on $(-\infty, 0)$, $\tilde{W}$ has a (unique) holomorphic extension to the upper half plane Δ_+. This extension is in the Hardy class H_1 on Δ_+, and so admits a canonical optimal-residual ("outer-inner") factorization

$$\tilde{W}(z) = \Phi(z)\Psi(z), \qquad z \in \Delta_+$$

where

$$(20.4) \quad \Phi(z) = \exp\left\{\frac{1}{\pi i} \int_{-\infty}^{\infty} \frac{\lambda z + 1}{\lambda - z} \frac{\log|\,\tilde{W}(\lambda)\,|}{1+\lambda^2} d\lambda\right\}, \qquad z \in \Delta_+$$

and Ψ is itself factorable into a Blaschke product, a factor, $\exp(i(\alpha + az))$ ($a \geq 0$, α real), and an integral akin to (20.4) but with respect to a purely singular and nonpositive measure over $(-\infty, \infty)$. In case $\Psi(z) = 1$, i.e. $\tilde{W} = \Phi$, the filter W_* is said to be of *minimum phase type*. The weighting for a filter of this type is easier to retrieve from input-output data than that of an arbitrary causal filter W_*, cf. §21.

In many cases of practical interest $\tilde{W}$ is a rational function (cf. §15). A second major problem in filter theory is that of *design* or *synthesis*: given a rational function $\tilde{W}$ satisfying (20.3), to synthesize from so-called lumped, passive, electrical elements (resistances, capacitances, inductances) the filter W_*. The problem of design is fairly well understood when the filter is linear. Wiener made some compelling suggestions as to its solution for nonlinear filters (§28).

21. Periodic and pulse inputs. To understand Wiener's work on filter inputs (§§22–25) we must review briefly the prior theory of purely periodic (in particular, sinusoidal) inputs and pulse inputs.

For the filter W_* we see at once that if $f(t) = e^{-i\lambda t}$, then

$$(21.1) \qquad g(t) = \sqrt{(2\pi)}\,\tilde{W}(\lambda)e^{-i\lambda t} = \sqrt{(2\pi)}\,\tilde{W}(\lambda)f(t).$$

Thus the response to a sinusoidal input is a sinusoidal signal of the same frequency $-\lambda$ but with a complex amplitude $\sqrt{(2\pi)}\,\tilde{W}(\lambda)$. The function which gives the amplitude modification, viz. $\sqrt{(2\pi)}\,\tilde{W}$, is called the *frequency response function* of the filter.[29] By successively plugging in sinusoidal inputs of different frequencies $-\lambda$ into the filter and observing the output amplitudes $\sqrt{(2\pi)}\tilde{W}(\lambda)$, we can get $\tilde{W}$ and thence W.

Again for the filter $W*$, since $W \in L_1(-\infty, \infty)$, it follows that if f is in $L_2(-\infty, \infty)$, i.e., f is an energetic but powerless pulse, then $g = W * f$ is also in $L_2(-\infty, \infty)$; moreover,

$$(21.2) \qquad \tilde{g}(\lambda) = \sqrt{(2\pi)}\tilde{W}(\lambda)\tilde{f}(\lambda).$$

Thus $\tilde{W}$ can be retrieved from a knowledge of the Fourier-Plancherel transforms $\tilde{f}$, $\tilde{g}$ of the input and output signal. Now cf. (1.1), (4.1),

$$(21.3) \qquad \begin{aligned} |\tilde{f}(\lambda)|^2 &= \textit{the energy density of f at frequency } -\lambda, \\ |\tilde{g}(\lambda)|^2 &= \textit{the energy density of g at frequency } -\lambda, \end{aligned}$$

and if T_t is the translation operator, then

$$(21.4) \qquad \int_{-\infty}^{\infty} e^{it\lambda}(T_t g, f)\,dt = \tilde{g}(\lambda)\overline{\tilde{f}(\lambda)} = \sqrt{(2\pi)}\tilde{W}(\lambda)|\tilde{f}(\lambda)|^2.$$

Since the functions f, g can be measured, the "output-input covariances" $(T_t g, f)$ for different lags t can be computed digitally, or analogically (by apparatus involving delay devices, square law rectifiers and integrators). Hence in principle we can find the cross output-input energy density (21.4). On dividing this by the input energy density (21.3) which again is computable from the observed signal f, we can find $\tilde{W}(\lambda)$ from (21.4).

In case the filter $W*$ is causal and of minimum phase type, the (unique) extension of $\tilde{W}$ in the upper half plane Δ_+ is given (cf. (20.4)) by

$$\tilde{W}(z) = \exp\left\{\frac{1}{\pi i}\int_{-\infty}^{\infty} \frac{\lambda z + 1}{\lambda - z} \frac{\log|W(\lambda)|}{1 + \lambda^2}\,dx\right\}, \qquad z \in \Delta_+$$

$$(21.5) \qquad = \exp\left[\frac{1}{2\pi i}\int_{-\infty}^{\infty} \frac{\lambda z + 1}{\lambda - z} \frac{1}{1 + \lambda^2}\log\left\{\frac{1}{2\pi}\frac{|\tilde{g}(\lambda)|^2}{|\tilde{f}(\lambda)|^2}\right\}d\lambda\right],$$

$$z \in \Delta_+, \text{ by } (21.2).$$

Thus in principle $\tilde{W}$ can be found merely from a knowledge of the input and output energy densities (21.3) without recourse to the cross

---

[29] More specifically, for an electric filter it is called the *impedance* or *admittance* or *voltage ratio*, according as $f = i$ & $g = v$, or $f = v$ & $g = i$, or $f = v$ & $g = v$, where i is the *current* and v is the *voltage*. We have defined this function to be $\sqrt{(2\pi)}\tilde{W}$ (and not $\sqrt{(2\pi)}\hat{W}$) in order that it be holomorphic in the upper (and not lower) half plane for a causal filter.

energy density (21.4). This is one of the merits of minimum phase filters.

A fictitious, limiting case of a pulse f as the support dwindles to a single point and the energy becomes infinite is the Dirac δ "function" supported at $\{0\}$ and such that $\delta(0) = \infty$. One assumes that $\int_{-\infty}^{\infty} |\delta(t)| \, dt = 1$, $\int_{-\infty}^{\infty} |\delta(t)|^2 dt = \infty$, so that δ is an *impact of "unit impulse" but infinite energy*. Formally, the response of the filter $W*$ to input δ is seen to be W:

$$(21.6) \qquad g(t) = (W * \delta)(t) = W(t), \qquad t \in (-\infty, \infty).$$

g will have finite energy if $W \in L_2(-\infty, \infty)$, i.e. $W \in L_1(-\infty, \infty) \cap L_2(-\infty, \infty)$. In this case

$$(21.7) \qquad \textit{the energy density of g at frequency } -\lambda = |W(\lambda)|^2.$$

The practical importance of (21.6) and (21.7) stems from the fact that as f approaches δ, g approaches W. Since approximations to δ are physically realizable by means of sharp, almost instantaneous impacts, we can find good approximations to W by subjecting our filter to such impacts and measuring its response.

22. **Inputs of the class** $\mathcal{S}$. One of Wiener's main contributions to linear filter theory was to realize the importance of admitting as inputs arbitrary signals of the class $\mathcal{S}$ (§2) and to extend the theory of L_2- and periodic signals (§21) to cover this more general case. His main theorem is that if

$$(22.1) \quad tW(t) \in L_1(-\infty, \infty), \quad (1 + |t|)W(t) \in L_2(-\infty, \infty),^{30}$$

then the response of the filter $W*$ to a signal f in $\mathcal{S}$ is a signal $g \in \mathcal{S}'$, and the spectral distributions S_f, S_g of f, g are related by

$$(22.2) \qquad S_g(\lambda) = 2\pi \int_{-\infty}^{\lambda} |W(u)|^2 dS_f(u), \qquad \lambda \in (-\infty, \infty)$$

[**FI**, Lemma 29$_6$, p. 173; Theorem 30, p. 178].

By a straightforward analysis Wiener showed that $g \in \mathcal{S}$ when $f \in \mathcal{S}$, but he had to appeal to the condition (2.8) of §2 on the "quadratic variation of s" to prove that g is actually in $\mathcal{S}'$. Latent in his work are results on the output-input covariance and spectral distribution ϕ_{gf}, S_{gf}:

$$\phi_{gf}(t) = (W * \phi_f)(t), \qquad t \in (-\infty, \infty),$$

$$(22.3)$$

$$S_{gf}(\lambda) = \sqrt{(2\pi)} \int_{-\infty}^{\lambda} \tilde{W}(u) dS_f(u), \qquad \lambda \in (-\infty, \infty).$$

---

[30] From which it follows readily that W and the two functions just mentioned are in fact in $L_1(-\infty, \infty) \cap L_2(-\infty, \infty)$.

Let the input signal f have an absolutely continuous spectral distribution S_f. It then follows that the spectra S_{gf}, S_g are also absolutely continuous, and (22.3), (22.2) reduce to

$$(22.4)\quad S'_{gf}(\lambda) = \sqrt{(2\pi)}\tilde{W}(\lambda)S'_f(\lambda), \qquad S'_g(\lambda) = 2\pi\,|\,\tilde{W}(\lambda)\,|^2 S'_f(\lambda).$$

The first of these equations corresponds to (21.2). Corresponding to (21.3) we now have, cf. (4.2) et seq.,

$$(22.5)\quad \begin{aligned} &S'_f(\lambda)/\sqrt{(2\pi)} = \textit{the power density of } f \textit{ at frequency } -\lambda.\\ &S'_g(\lambda)/\sqrt{(2\pi)} = \textit{the power density of } g \textit{ at frequency } -\lambda. \end{aligned}$$

As in the case of signals f, $g \in L_2(-\infty, \infty)$ (§21) ,we can find the output-input covariance $\phi_{gf}(t)$ for different lags t from observed data either by digital computation or by analogical devices, and in principle obtain the output-input power density $S'_{gf}(\lambda)$. On dividing this by the input power-density $S'_f(\lambda)$ we will get $\tilde{W}(\lambda)$, cf. (22.4). In case the filter $W*$ is causal and of minimum phase type, we can as in §21 retrieve $\tilde{W}$ from a knowledge of the power densities (22.5) alone without recourse to the cross power density S'_{gf}.

We see that Wiener was able to obtain for filters $W*$ a rather complete extension of the classical theory of L_2-inputs to S-inputs with absolutely continuous spectra.

Corresponding to the fictitious δ signal of the L_2-theory (§21) we now have the so-called unit *white noise signal* w characterized by the properties

$$\phi_w(t) = \delta(t), \qquad S'_w(\lambda) = 1/\sqrt{(2\pi)}, \qquad t, \lambda \in (-\infty, \infty).$$

Since $S_w(\infty) - S_w(-\infty) = \infty$, the signal w has infinite power. We may regard w as the fictitious limiting case of a signal f in S with a flat spectral density, the support of which swells to fill up the entire interval $(-\infty, \infty)$. Formally, the response g of the filter $W*$ to the input w satisfies

$$(22.6)\quad \phi_g(t) = W(t), \quad S'_{gw}(\lambda) = \tilde{W}(\lambda), \quad S'_g(\lambda) = \sqrt{(2\pi)}\,|\,\tilde{W}(\lambda)\,|^2.$$

On comparing this with (22.5) and (21.7) we can conclude that:

$$(22.7)\quad \begin{aligned} &\textit{If } W \textit{ satisfies (22.1), the power density function of the response}\\ &\textit{of the filter } W* \textit{ to the unit white noise input } w \textit{ is equal to the}\\ &\textit{energy density function of its response to the input } \delta, \textit{ i.e. to an}\\ &\textit{impact of unit impulse.} \end{aligned}$$

This is Wiener's result, being just a clearer rendition of his more cryptic formulation: *"the response of a linear resonator to a unit*

Brownian motion input has the same distribution of power in frequency that its response to a single instantaneous pulse will have as a distribution of energy in frequency" [**TS**, p. 50, **GHA**, p. 116]. We shall discuss its precise interpretation in §23.

23. **Brownian motion inputs.** The pseudo-concepts of the impact signal δ of infinite energy and of the white noise signal w of infinite power (§§21, 22) can be explicated in terms of "distributions" à la L. Schwartz. Integration by parts is an important tool in this explication. Such use of partial integration actually goes back to Wiener's early work on Brownian motion [29]. In [**GHA**, §13] Wiener adopted this method to convert pseudo-assertions concerning the response of the filter $W*$ to white noise into bona fide assertions about the response of the filter $-dW*$ to a Brownian motion input.

To explain Wiener's approach, let us first see how the pseudo-assertion, (21.6) about the response of $W*$ to δ, can be reinterpreted as a result concerning the response of $-dW*$ to the *Heaviside input* $\chi_{[0,\infty)}$. This input invokes from the filter $-dW*$ the response

$$g(t) = -\int_{-\infty}^{\infty} d_\tau W(t-\tau)\,\chi_{[0,\infty)}(t) = -\int_0^\infty d_\tau W(t-\tau) = W(t).$$

Here we assume that W is of bounded variation on $(-\infty,\infty)$ and is in $L_2(-\infty,\infty)$, so that $W(-\infty)=0$. Thus, *the response of the filter* $-dW*$ *to the Heaviside input* $\chi_{[0,\infty)}$ *is* W. This statement would be the Wienerian explication of the pseudo-statement: the response of the filter $W*$ to δ is W.

Wiener handled the white noise input in a similar way. Let $\{x(t,\alpha),\ t\in(-\infty,\infty),\ \alpha\in[0,1)\}$ be a (separable) BMSP. The response g_α of the filter $-dW*$ to a Brownian motion $x(\cdot,\alpha)$ is given by

$$(23.1)\quad g_\alpha(t) = -\int_{-\infty}^{\infty} d_\tau W(t-\tau)\cdot x(\tau,\alpha) \qquad t\in(-\infty,\infty).$$

To ensure the existence of the last integral Wiener assumed [**GHA**, p. 225] that

$$(23.2)\quad W(t) = \frac{K(t)}{\sqrt{(1+t^2)}}, \qquad K \text{ of bounded variation on } (-\infty,\infty).$$

Obviously $W\in L_2(-\infty,\infty)$ and is itself of bounded variation on $(-\infty,\infty)$. Since for almost all α, $x(\cdot,\alpha)$ is continuous on $(-\infty,\infty)$ and $|x(t,\alpha)|\leq 2\sqrt{(|t|\ \log |t|)}$, $|t|$ large, the integral in (23.1) exists for almost all α. Now we can show by partial integration that there exists a fixed set $N\subseteq[0,1]$ of zero Lebesgue measure such that for each $t \in (-\infty, \infty)$, there is a version $y(t, \cdot)$ of the stochastic integral

$$(23.3) \qquad \int_{-\infty}^{\infty} W(t - \tau)dx(\tau, \cdot)$$

such that

$$(23.4) \quad y(t, \alpha) = g_\alpha(t), \qquad \alpha \in [0, 1] - N, \quad t \in (-\infty, \infty).[31]$$

Let us observe that the stochastic integral (23.3) is precisely the one defined in (5.4) above. Our present $y(t, \cdot)$ is just a well chosen member of the equivalence class $y(t,)$ of (5.4). It follows from (23.4), (22.5) and (5.5) that

(23.5) *for almost all α the response of the filter $-dW*$ to the Brownian motions $x(\cdot, \alpha)$ belongs to S' and has the (same) power density function $|\tilde{W}(-\cdot)|^2$.*

With Wiener we may consider (23.5) as the explication of the pseudo-statement that $|\tilde{W}(-\cdot)|^2$ is the power-density function of the response of the filter $W*$ to the white noise input w. The pseudo statement (22.7) may likewise be given the precise rendering:

(23.6) *If W satisfies (23.2), then the power density function of the response of the filter $-dW*$ to almost all Brownian motions is equal to the energy density function of its response to the Heaviside input $\chi_{[0,\infty)}$.*

24. On harmonic analysis and linearity. From the preceding account it is apparent that harmonic analysis, classical and generalized, is a valuable tool in the study of time-invariant, linear filters. What, however, is the precise connection between harmonic analysis and linearity? Wiener had a clear understanding of this question, and this fact was crucial in the evolution of his ideas on nonlinear filters. Unfortunately his writings on the question are somewhat vague [204, pp. viii–x, 51–53, **NPRT**, p. 90]. We shall therefore make some clarifying remarks before proceeding to his work on nonlinear filters.

Let G be any (additive) abelian Haar-measured group and $\hat{G}$ be its character group. Let S_t be the *translation operator* induced by $t \in G$ on the space of complex-valued functions on G:

$$\{S_t(f)\}(x) = f(x + t), \qquad t, x \in G.$$

Let T be any linear operator on some subspace, e.g. $L_2(G)$, of such

---

[31] The proof of (23.4) involves some delicate questions concerning measurability and the behavior of mobile sets of zero-measure, which Wiener seems to have slurred over. The writer is grateful to Professor T. Hida for assistance in settling these questions.

functions into itself, which commutes with each S_t. Then each character α of G is an eigenfunction of T:

$$T(\alpha) = Y(\alpha) \cdot \alpha,$$

where $Y(\alpha) = \{T(\alpha)\}(0)$. It follows that if f has a "Fourier expansion" in terms of the characters

$$f(x) = \sum_j c_j \alpha_j(x), \qquad x \in G, \quad \alpha_j \in \hat{G},$$

then

$$\{T(f)\}(x) = \sum_j c_j Y(\alpha_j) \cdot \alpha_j(x).$$

Hence $T(f)$ can be found, if we know the "Fourier coefficients" c_j of f and the "transfer function" (or "impedance") Y of T. In short, T is completely determined by the function Y on $\hat{G}$, i.e. in filter terminology, by T's "response" to the "character inputs" α. "Harmonic analysis" thus appears as the natural tool for the study of linear operators which commute with the translations S_t.

In the case of time-invariant linear filters, G and therefore $\hat{G}$ are of course the additive group of real numbers, and the harmonic analysis just alluded to becomes the classical harmonic analysis, or for robust, nonperiodic signals Wiener's generalized harmonic analysis, the character-inputs being sinusoidals. The transfer function Y becomes the frequency response function, i.e. the Fourier transform of the weighting function (§§20, 21).

Next let T be a *nonlinear* operator on some subspace of complex-valued functions on G, again commuting with the translations S_t. Then in general T is not determined by its action on the characters of G. Consequently "character-inputs" (i.e. sinusoidals) do not play any intrinsic role, and harmonic analysis ceases to be especially significant. An important problem is to find the right substitutes: an appropriate probe, and an appropriate analysis of the response of the filter to such probes. With encouragement from Vannevar Bush, Wiener kept pondering on this question starting in the twenties.

25. **The probe for a nonlinear filter.** Among Wiener's many statements on the question of a probe, perhaps the most succinct and lucid is that occurring in his posthumously published book [211, p. 34–]. He writes:

> The output of a transducer excited by a given input message is a message that depends at the same time on the input message and on the transducer itself. Under the most usual circumstances, a transducer is a mode of transforming messages, and our attention is drawn to the output message as a transformation of the input message. However, there are circumstances, and these chiefly arise when the input message carries a minimum of information, when we may conceive the information of the out-

put message as arising chiefly from the transducer itself. No input message may be conceived as containing less information than the random flow of electrons constituting the shot effect. Thus the output of a transducer stimulated by a random shot effect may be conceived as a message embodying the action of the transducer.

As a matter of fact, it embodies the action of the transducer for any possible input message. This is owing to the fact that over a finite time, there is a finite (though small) possibility that the shot effect will simulate any possible message within any given finite degree of accuracy. . . . That is, if we know how a transducer will respond to a shot-effect input, we know ipso facto how it will respond to any input.

The belief that the nature of a filter can be found by studying its response to the random stimulation it receives from its environment—by psychoanalysis so-to-speak rather than by lobotomy—came rather early in Wiener's development. Thus in [GHA, p. 215] we read: "Imagine a resonator—say a sea-shell—struck by a purely chaotic sequence of acoustical impulses. It will yield a response which still has a statistical element in it, but in which the selective properties of the resonator will have accentuated certain frequencies at the expense of others."

The shot effect of which Wiener speaks is to be regarded as the physical realization of a white noise signal $w(\cdot, \alpha)$ or more accurately of a Brownian incrementary signal $dx(\cdot, \alpha)$, (cf. §§22, 23). Our experiences with the Taylor expansion of ordinary (nonlinear) functions on $\mathfrak{R}^n$ suggest that the response of our filter to a white noise signal should be expressible as a linear combination of multiple stochastic integrals

$$(25.1) \quad \int_{-\infty}^{t} K(s)dx(s, \alpha), \qquad \int_{-\infty}^{t}\int_{-\infty}^{t} K(s, \sigma)dx(s, \alpha)dx(\sigma, \alpha), \text{ etc.}$$

Wiener had already encountered such integrals in his researches on the *homogeneous chaos* [108], [128], the motivation for which lay in ergodic theory and statistical mechanics. This work, he now found, had a direct bearing on the problem of nonlinear filters, cf. §26.

26. **Fourier-Hermite series of a function in $L_2[0, 1]$.** Wiener came upon this subject in 1938 in his attempts to free ergodic theory from dependence on one-one point transformations [108]. In the course of this work he originated the idea of a *homogeneous chaos over $\mathfrak{R}^n$*, i.e. a function F on $\mathfrak{B}_n \otimes [0, 1]$, ($\mathfrak{B}_n$ = the family of Borel subsets of $\mathfrak{R}^n$) such that for each $\alpha \in [0, 1]$, $F(\cdot, \alpha)$ is finitely additive on $\mathfrak{B}_n$, and for each $t \in \mathfrak{R}^n$, each $S \in B_1$, and each $B \in \mathfrak{B}_n$, the sets

$$\{\alpha: F(B + \{t\}, \alpha) \in S\}, \quad \{\alpha: F(B, \alpha) \in S\}$$

have the same Lebesgue measure [108, §2]. The current term for such an F is: *stationary random measure*.

114   P. MASANI

The simplest example of a homogeneous chaos is the independently scattered Brownian measure over $\Re$, defined on $\mathcal{B}_1 \otimes [0, 1]$ by extending

$$F((a, b], \alpha) = x(b, \alpha) - x(a, \alpha), \qquad a < b,$$

where $\{x(t, \alpha),\ t \in \Re,\ \alpha \in [0, 1]\}$ is the BMSP. This F is called a 1-*dimensional pure chaos* [108, §6]. The multiple Wiener stochastic integrals yield *derived chaoses*; e.g. over $\Re^n$ we have the chaos

$$F(B, \alpha) = \int \cdots \int_B K(t)dx(t_1, \alpha) \cdots dx(t_n, \alpha),$$

due to $K \in L_2(\Re^n)$; here $t = (t_1, \cdots, t_n)$. Next, let

$$f(t, \alpha) = \int \cdots \int_{\Re^n} K(t - \tau)dx(\tau_1, \alpha) \cdots dx(\tau_n, \alpha)$$

and define

$$F(B, \alpha) = \int \cdots \int_B f(t, \alpha)dt_1 \cdots dt_n.$$

Then F is called an *nth-degree homogeneous polynomial chaos* [108, §9]. Such considerations led Wiener to the idea of a hierarchy of mutually orthogonal subspaces of $L_2[0, 1]$, [19, p. 37 et seq]. The following version of his theorem is an adaptation of Kakutani's (33, Theorem 1):

THEOREM 1. *Let $\mathcal{S}_0 = \Re$, and for $n \geq 1$, $\mathcal{S}_n$ be the subspace of symmetric functions in $L_2(\Re^n)$. Then for each $n \geq 0$, there is a linear isometry G_n on $\mathcal{S}_n$ into $L_2[0, 1]$ such that*

$$L_2[0, 1] = \sum_0^\infty G_n(\mathcal{S}_n), \qquad G_m(\mathcal{S}_m) \perp G_n(\mathcal{S}_n), \qquad m \neq n,$$

$$U_t\{G_n(\mathcal{S}_n)\} = G_n(\mathcal{S}_n),$$

where U_t is the unitary flow on $L_2[0, 1]$ induced by the flow of the BMSP, more accurately of white noise, on $([0, 1], Borel, Leb.)$, cf. (5.6).[32]

The expressions for G_0 and G_1 are easily found:

$$\{G_0(c)\}(\alpha) = c, \qquad c \in \Re,$$

(26.1)
$$\{G_1(\phi)\}(\alpha) = \int_{-\infty}^{\infty} \phi(t)dx(t, \alpha), \qquad \phi \in L_2(-\infty, \infty).$$

Those for G_2, G_3 etc. are best understood from the remarkable rela-

---

[32] Our G_n differs from Wiener's in that his $G_n(K_n, \alpha)$ is our $\sqrt{(n!)}\{G_n(K_n)\}(\alpha)$, K_n being in $L_2(\Re^n)$. Our G_n is Kakutani's W_n^{-1}.

tionship which subsists between the G_n and the orthonormal polynomials H_n, $n \geq 0$, of the $(0, 1)$ normal distribution over $\mathfrak{R}$:

$$\int_{-\infty}^{\infty} H_m(u) H_n(u) \frac{1}{\sqrt{(2\pi)}} e^{-u^2/2} du = \delta_{mn}.$$

The H_n are of course the *Hermite polynomials*

$$H_n(u) = (n!)^{-1/2} (-1)^n e^{u^2/2} D_u^n (e^{-u^2/2}), \qquad u \in \mathfrak{R}.$$

The relationship in question is that if $\{\phi_n,\ n \geq 0\}$ is any o.n. subset of $L_2(\mathfrak{R})$, and $p_0, \cdots, p_n$ are non-negative integers, then [**NPRT**, p. 94][33]

$$\left(\left(\sum_{i=0}^{m} p_i \right)! \right)^{1/2} G_{\Sigma_0^m p_i} \left\{ \overset{m}{\underset{i=0}{\mathsf{X}}} \phi_i^{(p_i)} \right\} = \prod_{i=0}^{m} (p_i!)^{1/2} H_{p_i} \{ G_1(\phi_i) \} \ \in \ L_2[0,1].$$

In view of this and Equation (26.1), we may recast Theorem 1 as a result on orthonormal expansions. The following enunciation of this is suggested by Cameron and Martin (8, Theorem 1).[34]

THEOREM 2 (Fourier-Hermite expansion). *Let* $f \in L_2[0, 1]$, $\{\phi_n,\ n \geq 0\}$ *be an o.n. basis for* $L_2(\mathfrak{R})$, *and let for all non-negative integers* $p_0, \cdots, p_m$,

$$a_{p_0, \cdots, p_n} = \left(f(\cdot),\ \prod_{j=0}^{m} H_{p_j} \left\{ \int_{-\infty}^{\infty} \phi_j(\tau) dx(\tau,\ \cdot) \right\} \right).$$

Then

$$f(\cdot) = \sum_{(p_0, \cdots, p_m)} a_{p_0, \cdots, p_m} \prod_{j=0}^{m} H_{p_j} \left\{ \int_{-\infty}^{\infty} \phi_j(\tau) dx(\tau,\ \cdot) \right\},$$

the convergence being in the $L_2[0, 1]$ *topology.*

The $a_{p_0, \cdots, p_m}$ are called the *Fourier-Hermite coefficients* of f, and the last series the *Fourier-Hermite series* of f relative to the basis

---

[33] For functions $f_{r_0}, \cdots, f_{r_m}$ on $\mathfrak{R}^{r_0}, \cdots, \mathfrak{R}^{r_m}$, respectively, we define on $\mathsf{X}_{i=0}^m f_{p_i}$ on $\mathfrak{R}^{\Sigma_0^m p_i}$ by

$$\left(\overset{m}{\underset{i=0}{\mathsf{X}}} f_{p_i} \right)(t_1, \cdots, t_{\Sigma_0^m p_i}) = \prod_{j=0}^{m} f_{p_j} \{ t_{\Sigma_0^{j-1} p_i + 1}, \cdots, t_{\Sigma_0^j p_i} \}.$$

We define $f^{(p)} = f \mathsf{x} \cdots \mathsf{x} f$ (p times).

[34] To get their Theorem 1 from our Theorem 2 we must, however, replace $t \in (-\infty, \infty)$ by $t \in [0, 1]$, and our $\alpha \in [0, 1]$ by the path $x(\cdot, \alpha) \in C_0[0, 1]$, where $C_0[0, 1]$ is the support of *Wiener measure*, i.e., the space of continuous functions $x(\cdot)$ on $[0, 1]$ such that $x(0) = 0$. It is outside our purview to report on the important role played by Wiener measure over $C_0[0, 1]$ in contemporary analysis. We refer to the work of Cameron and Martin, M. Kac, Feynman, K. Ito and others.

$\{\phi_n,\ n\geq 0\}$ of $L_2(\mathcal{R})$. This series plays an important role in Wiener's theory of nonlinear filters.

27. **The response of a nonlinear filter to white noise.** Let $g(\cdot,\ \alpha)$ be the response of our filter to the white noise signal $w(\cdot,\ \alpha),\ \alpha\in[0,\ 1]$. We assume first that the filter is *time-invariant*, i.e., if g is the response to the input f, then the translate $g(\cdot+h)$ is the response to the input $f(\cdot+h)$. Taking $f(\cdot)=w(\cdot,\ \alpha)$, and recalling that

$$(27.1) \qquad\qquad w(\tau+t,\ \alpha) = w(\tau,\ T_t\alpha),$$

where T_t is the flow of white noise over $([0,\ 1],\ \text{Borel, Leb.})$, cf. (5.6), it follows from the time-invariance of the filter that

$$(27.2) \qquad g(\tau+t,\ \alpha) = g(\tau,\ T_t\alpha), \qquad g(t,\ \alpha) = g(0,\ T_t\alpha).$$

We assume next that our filter is *stable* in the sense that

$$(27.3) \qquad\qquad g(0,\ \cdot) \in L_2[0,\ 1],$$

(and hence $g(t,\ \cdot)\in L_2[0,\ 1]$). Wiener's remarks [**NPRT**, p. 89] as to how this condition restricts the filter are not clear to the writer. (The nexus between the ideas under consideration and the subject of non-linear oscillations needs investigation.)

By (27.3) and §26, Theorem 2, $g(0,\ \cdot)$ has a Fourier-Hermite development relative to any o.n. basis $\{\phi_n,\ n\geq 0\}$ of $L_2(\mathcal{R})$. From (27.1) and (27.2) it easily follows that to get the Fourier-Hermite series for $g(t,\ \cdot)$ we have only to replace the functions $\phi_n(\cdot)$ by their translates $\phi_n(\cdot-t)$. Thus

$$(27.4)\ \ g(t,\ \cdot) = \sum_{(p_0,\cdots,p_m)} a_{p_0,\cdots,p_m} \prod_{j=0}^{m} H_{p_j}\left\{\int_{-\infty}^{\infty}\phi_j(\tau-t)dx(\tau,\ \cdot)\right\},$$

$$t\in\mathcal{R}.$$

Finally, we assume that our filter is *causal*, i.e., if $g_1,\ g_2$ are its responses to inputs $f_1,\ f_2$, then for each t,

$$f_1 = f_2 \text{ on } (-\infty,\ t) \Rightarrow g_1 = g_2 \text{ on } (-\infty,\ t).$$

It then follows that in (27.4) the upper terminus of the integrals must be t and not ∞. In effect the functions ϕ_n must be supported on $(-\infty,\ 0]$, i.e., $\{\phi_n,\ n\geq 0\}$ must be an o.n. basis for $L_2(-\infty,\ 0]$. To conclude, *each time-invariant, stable, causal filter can be characterized in terms of any o.n. basis $\{\phi_n,\ n\geq 0\}$ of $L_2(-\infty,\ 0]$ by the system of coefficients $a_{p_0,\cdots,p_m}$ occurring in the Fourier-Hermite expansion (27.4) of its response at any given moment to white noise.*

By pursuing Wiener's heuristic idea quoted in §25 one should be able to prove that the coefficients $a_{p_0,\cdots,p_m}$ suffice to determine the response of our filter to *any* reasonable input, not just white noise. One would then be able to assert that *to "know" the filter is to know the coefficients* $a_{p_0,\dots,p_m}$ *relative to any o.n. basis of* $L_2(-\infty, 0]$. Such a proof does not seem to have been carried out, however, and hence it is not clear (to the writer at least) if Wiener's theory is equivalent to the others, which have been suggested, based on less random probes, cf. e.g. D. Gabor (23).

28. **Synthesis and analysis of nonlinear filters.** In the *synthesis problem* the function $g(\cdot, \alpha)$ is prescribed, and we have to design a filter ("white box") for which the response to white noise $w(\cdot, \alpha)$ is this $g(\cdot, \alpha)$. In the *analysis problem* we are given a "black box" the response of which to white noise is observable, and we are asked to determine its characterizing coefficients $a_{p_0,\dots,p_m}$ relative to some o.n. basis of $L_2(-\infty, 0]$, cf. (27.4).

Wiener solved both problems. In his solution a crucial step was his selection of the functions $L_n(-\cdot)$, where L_n is the *nth Laguerre function*:

$$L_n(t) = e^{-t} i \sum_{k=0}^{n} (-1)^{k+1} 2^{k+1/2} (k!)^{-1} \binom{n}{k} t^k, \qquad t \in [0, \infty), \quad n \geq 0,$$

for the o.n. basis for $L_2(-\infty, 0]$. Letting $\phi_j(t) = L_j(-t)$ and writing $w_\alpha(t)$ instead of $w(t, \alpha)$ for convenience, we see at once that

$$\int_{-\infty}^{\infty} \phi_j(\tau - t) dx(\tau, \alpha) = \int_{-\infty}^{t} L_j(t - \tau) w_\alpha(\tau) d\tau = (L_j * w_\alpha)(t);$$

i.e., the term on the left is the response at instant t of the *linear filter* L_j* to the white noise signal w_α, cf. (20.2), (20.2'). Wiener usually refers to $(L_j*w_\alpha)(t)$ as the (jth) "Laguerre coefficient of the past of the input." It follows from (27.4) that relative to this Laguerre basis

$$(28.1) \qquad g(t, \alpha) = \sum_{(p_0,\dots,p_m)} a_{p_0,\dots,p_m} \prod_{j=0}^{m} H_{p_j}\{(L_j * w_\alpha)(t)\},$$

and that *our filter ("black" or "white") can be characterized by the system of Laguerre-Hermite coefficients* $a_{p_0,\dots,p_m}$.

Now in 1928 Wiener and Y. W. Lee showed how the linear filter L_j* ($j \geq 0$) can be built from lumped, passive electrical elements— the so-called *Laguerre networks*. As Wiener saw and as we shall now indicate, this possibility along with (28.1) permits a theoretical solution of both the synthesis and analysis problems.

In the synthesis problem the Laguerre-Hermite coefficients $a_{p_0,\dots,p_m}$ are given. The equation (28.1) shows that we can approximate to a filter with these coefficients to any desired degree (with the RMS error criterion) by suitably cascading a large number of Laguerre networks with square law rectifiers (multipliers), scale amplifiers and summing circuits. For the analysis problem we find from (28.1) with $t=0$, that

$$a_{p_0,\ldots,p_m} = \int_0^1 g(0,\,\alpha) \cdot \prod_{j=0}^m H_{p_j}\{(L_j * w_\alpha)(0)\}\,d\alpha.$$

Since the Brownian flow $(T_t,\ t \in \mathcal{R})$ is ergodic, it follows from the Ergodic Theorem and the relations

$$g(0,\,T_t\alpha) = g(t,\,\alpha), \qquad w(0,\,T_t\alpha) = w(t,\,\alpha)$$

that for almost all α

$$a_{p_0,\ldots,p_m} = \lim_{\tau \to \infty} \frac{1}{\tau} \int_{-\tau}^0 g(t,\,\alpha) \prod_{j=0}^m H_{p_j}\{(L_j * w_\alpha)(t)\}\,dt.$$

Now the second factor in the integrand is the response to the white noise signal w_α of a "white box," easily constructible from Laguerre networks, square law rectifiers and summing circuits, and the first factor $g(t,\,\alpha)$ is the "observed" response of our black box to this signal. Thus, by simultaneously bombarding our black box and the white box with the same white noise signal, and passing the outputs through a square law rectifier and integrating device we can approximate to the coefficients $a_{r_0,\ldots,p_m}$. In this way both the synthesis and analysis problems can be tackled, in principle at least.

29. **Reproduction, learning, self-organization.** We can comment only very briefly on the bearing of Wiener's theory of filters on the larger questions of the reproducing, learning and self-organizing abilities of natural systems.

(i) Wiener realized that his solution of the analysis and synthesis problems for filters provided a theorita[35] for the regeneration of filters. In the block diagram on p. 119 W is a white box comprising Laguerre networks, square law rectifiers, scale magnifiers and summing circuits, wired according to the scheme of (28.1). All parameters are fixed except those (inductances) which determine the coefficients $a_{p_0,\ldots,p_m}$. W has a number of terminals into which signals controlling these parameters can be fed. Thus W is potentially capable of performing the operation of any time-invariant, stable, causal filter.

Now suppose that we wish to create an operative image of a black box B. We can obtain its $(p_0,\cdots,p_m)$th Laguerre-Hermite coefficient by feeding the same white noise into B and the (known) $(p_0,\cdots,p_m)$th white box, and averaging the product of the outputs (cf. §28, end). By feeding these averages into the coefficient-control-

---

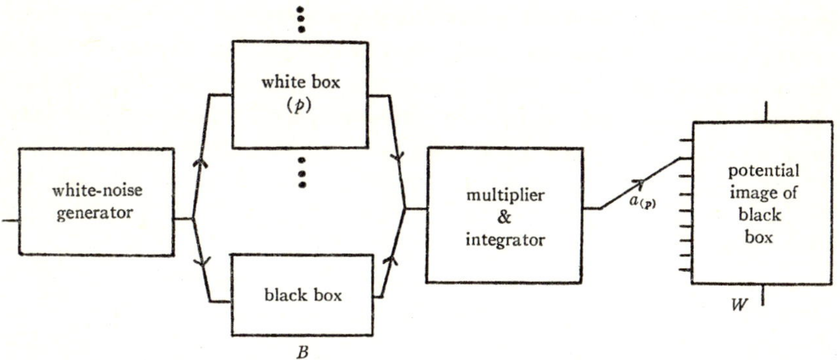

ling terminals of W (cf. Fig.), we can transform W into an actual operative image of B. Thus a machine can make another in its own likeness.

(ii) A *designed* or *purposive filter* is one whose response g to an input f approximates to $T(f)$, where T is some preassigned (well-defined) operator. The difference $g-T(f)$ between the actual and ideal response, or some norm thereof, is called its *error of performance*. A purposive filter is said to *learn*, if in the course of its operation the error of performance decreases. The existence of such filters is now well recognized.

Wiener visualized a learning filter as a system comprising a performing filter I coupled by feedback to a nonlinear filter II, [204, p. 173; 211, pp. 14, 20–21]. I carries out the routine of transforming input f into output g. II keeps a record of past inputs, outputs, and errors of performance of I, and has devices for the re-evaluation of the parameters governing this performance. Periodically, the system takes "time-out" to make this re-evaluation. The results are automatically fed back to I, and the routine resumed with improved efficiency. For instance, the component II of a learning anti-aircraft battery would record the long-time trajectories of incoming planes, and compute therefrom estimates of the covariances of the hypothetical underlying SP. For this it would have to include nonlinear devices such as square law rectifiers. When these improved estimates of the covariances are fed into component I of the battery, the new g's produced by I will be closer to the ideal predictor $T(f)$ than before, i.e. the performance will improve.

Wiener compared this ability of a filter to fulfil its own purpose better by appropriately modifying its responses in the light of relevant external realities to *biological adaptation* (ontogenetic learning). He compared the propagation of such "self-educated" filters with racial or phylogenetic learning.

(iii) Wiener supplemented the ideas just recounted to explain the anti-entropic, self-organizing activity of other, less blue-printed, natural systems. One such activity is the phenomenon of *gating* in the nervous system of animals [204, pp. 197–198]. This function, vitally important for the efficient use of nervous tissue, suggests the presence of a clock in the brain. How would such a clock survive the continual random bombardment it gets from its surroundings? Wiener's answer lay in his theory of *entrainment* or *attraction of frequencies*. The following is a brief outline of the mathematical side of this [NPRT, Lecture 8].

Consider an assembly of nonlinearly coupled oscillators, which in ideal isolation vibrate harmonically at the (same) fixed frequency ω. Because of the coupling the undisturbed oscillation $e^{i\omega t}$ gives way to

$$(29.1) \qquad y(t, \alpha) = e^{i\omega t} f(t, \alpha)$$

where

$$(29.2) \quad f(t, \alpha) = \exp \left\{ i\epsilon \iint_{\mathcal{R}^2} K(t + \tau_1, t + \tau_2) dx(\tau_1, \alpha) dx(\tau_2, \alpha) \right\},$$

$$\epsilon > 0,$$

i.e., the "carrier" $e^{i\omega t}$ is "quadratically frequency-modulated by white noise."[36] Actually for computational convenience Wiener took $\omega = 0$ in (29.1) (heterodyned frequency). He also assumed that K is symmetric and that ϵ is so small that in the expansion

$$f(t, \alpha) = 1 + i\epsilon \iint_{\mathcal{R}^2} K(t + \tau_1, t + \tau_2) dx(\tau_1, \alpha) + O(\epsilon^2)$$

the terms $O(\epsilon^2)$ can be neglected. He then showed that the spectral distribution F of the SP $y(\cdot, \cdot)$ has a sharp line at the (heterodyned) frequency 0, as well as an absolutely continuous part with density

$$(29.3) \qquad F'(\lambda) = \frac{\epsilon^2}{\sqrt{(2\pi)}} \int_{-\infty}^{\infty} | Q(u, \lambda - u) |^2 du,$$

where Q is the direct Fourier-Plancherel transform of K in $L_2(\mathcal{R}^2)$. Actually the sharp line is smeared by the Doppler effect due to thermal noise, the resulting distribution being of Cauchy type [189]. The upshot is that the SP $y(\cdot, \cdot)$ has a spectral density which is the sum of (29.3) and a Cauchy density.

Now Wiener was guided by the case of an electric power generating system in which, through negative feedback, a number of alternators can maintain a sharp frequency despite variations of load. The fre-

---

[36] This is of course a simplification; we should also allow linear, cubic and higher order integrals in the { } in (29.2), and determine their contribution.

quencies "attract," i.e., the coupling is such that slow alternators are speeded up and fast ones slowed down. This led him to stipulate that

$$(29.4) \qquad K(-t_1, -t_2) = - K(t_1, t_2)$$

—a kind of reaction principle. (29.4) entails that $Q(u, -u) = 0$ for all u; whence $F'(0) = 0$. Putting the pieces together we find that the process $y(\cdot, \cdot)$ in (29.1) has a spectral density with a sharp peak at ω, with a dip and a small hill on either side. This profile is actually observed in the empirical spectrum of brain wave encephalographs.

Thus by dint of nonlinear couplings, the oscillators in an assembly can maintain narrow frequency bands, despite the presence of noise. Just as E. Hille saw semi-groups where the less initiated saw none, so Wiener detected entrainment in very diverse situations, e.g. the diurnal rhythm in many animals, the flashing of fire-flies in unison, the lumping in the periods of the asteroids, the breakdown of the earlier models of electra and comet airplanes. In [210], which was perhaps his last mathematical paper, Wiener attempted to show that a Hamiltonian or even low dissipative dynamical system, excited by random turbulence, would under certain circumstances generate non-linear oscillations confined to narrow frequency bands. He even felt that quantum phenomena could be explained in this way. But the systematic discussion of Wiener's ideas on quantum theory, statistical mechanics and brain waves is beyond our scope.

30. **Epilogue.** Much remains to be done, especially at the non-linear level, to set up Wiener's work on filters as a rigorous mathematical discipline. There are points of contact between his ideas and those conceived in the theories of nonlinear oscillations, of higher order spectra and general automata, which obviously need systematic exploration.

The ideas underlying Wiener's theory are very far reaching, and touch upon the very concepts of existence and progress. A purposive filter is able to produce a local zone of organization by absorbing energy and information from its environment, and maintaining homeostasis by feedback. But every filter must in the course of time wither away. As Wiener remarked, "The paradox of homeostasis is that it always breaks down in the end."[37] The value of existence should thus be gauged not in terms of sheer survival but in terms of the necessarily fugitive pursuit of anti-entropic activity. In [177, pp. 324–325] Wiener describes this predicament eloquently in relation to his own existence. He concludes: "The declaration of our own nature and the attempt to build up an enclave of organization in the

---

[37] Quotation from an unpublished manuscript.

face of nature's overwhelming tendency to disorder is an insolence against the gods and the iron necessity that they impose. Here lies tragedy, but here lies glory too."

But there is another aspect to this matter with which Wiener was also concerned. In the life of a highly organized filter may come moments when the fulfillment of its phylogenetic responsibilities might necessitate its own destruction. For phylogenetic survival such filters must learn a new art—that of sacrifice. The practice and theory of this art is religion, and mankind has always associated its highest manifestation with martyrdom. Can a theory of sacrifice be formulated within the framework of an enlarged theory of filters? Wiener mused on this and related questions in his Terry Lectures at Yale and in his posthumously published book [211]. Some of his ideas are profound, and suggest a way to free modern thought from the lurking inconsistency between the scientific and religious positions, which A. N. Whitehead and other great thinkers have found so vitiating.[38]

In the course of his life Wiener tried to entrain many a filter to his point of view. His presence has been a great stimulus and a great challenge. One way we can express our affection and gratitude is to strive for an intellectual and moral climate in which others like him may arise. A more immediate task is to explore further the many ideas he has left behind. On the mathematical side, we have the problems of nonlinear innovations, of entrainment in relation to nonlinear oscillations, and the clarification of the notion of a stable filter, not to mention many others in areas like quantum mechanics, which we have not discussed.

Frequently Quoted Works of Norbert Wiener

GHA *Generalized harmonic analysis*, Acta Math. **55** (1930), 117–258.

FI *The Fourier integral and certain of its applications*, Cambridge Univ. Press, New York, 1933.

TS *Extrapolation, interpolation and smoothing of stationary time series with engineering applications*, The MIT Press, Cambridge, Mass.; Wiley, New York; Chapman & Hall, London, 1949.

NPRT *Nonlinear problems in random theory*, The MIT Press, Cambridge, Mass. and Wiley, New York, 1958.

Other References

1. K. Balagangadharan, *The prediction theory of stationary random distributions*, Mem. Coll. Sci. Univ. Kyoto Ser. A **33** (1960), 243–256.

2. G. Baxter, *An operator identity*, Pacific J. Math. **8** (1958), 649–663.

---

[38] A. N. Whitehead, *Science and the modern world*, Cambridge (1933), p. 94.

3. A. Beurling, *On two problems concerning linear transformations in Hilbert space*, Acta Math. 81 (1949), 239–255.

4. G. D. Birkhoff, *Proof of the ergodic theorem*, Proc. Nat. Acad. Sci. U.S.A. **17** (1931), 656–660.

5. S. Bochner, *Lectures on Fourier Integrals*, Annals of Mathematics Studies No. 42, Princeton, N.J., 1959.

6. ———, *Harmonic analysis and the theory of probability*, Univ. of Calif. Press, Berkeley, Calif., 1955.

7. H. W. Bode and C. E. Shannon, *Linear least square smoothing and prediction theory*, Proc. Inst. Elec. Radio Engrs **38** (1950), 417–425.

8. R. H. Cameron and W. T. Martin, *The orthogonal development of non-linear functionals in series of Fourier-Hermite functionals*, Ann. of Math. 48 (1947), 385–392.

9. J. Chover, *Conditions on the realization of prediction by measures*, Duke Math. J. 25 (1958), 305–310.

10. ———, *A theorem of integral transforms with an application to prediction theory*, J. Math. Mech. 8 (1959), 939–945.

11. H. Cramer, *On the representation of functions by certain Fourier integrals*, Trans. Amer. Math. Soc. 46 (1939), 191–201.

12. ———, *On the theory of stationary random processes*, Ann. of Math. 41 (1940), 215–230.

13. ———, *On the linear prediction problem for certain stochastic processes*, Ark. Mat. 4 (1959), 45–54.

14. ———, *On some classes of non-stationary stochastic processes*, Proc. Fourth Berkeley Symposium on Mathematical Statistics and Probability, Univ. of Calif. Press, Berkeley, Calif., 1960, 57–78.

15. ———, *On the structure of purely non-deterministic stochastic processes*, Ark. Mat. 4 (1961), 249–266.

16. S. Darlington, *Linear least-squares smoothing and prediction with applications*, Bell System Tech. J. 37 (1958), 1221–1294.

17. A. Devinatz, *The factorization of operator valued functions*, Ann. of Math. **73** (1961), 458–495.

18. C. L. Dolph and M. A. Woodbury, *On the relation between Green's functions and the covariances of certain stochastic processes and its application to unbiassed linear prediction*, Trans. Amer. Math. Soc. **72** (1952), 519–550.

19. J. L. Doob, *Time series and harmonic analysis*, Proc. Berkeley Symposium on Mathematical Statistics and Probability, Univ. of Calif. Press, Berkeley, Calif., 1949, 303–343.

20. ———, *Stochastic processes*, Wiley, New York; Chapman & Hall, London; 1953.

21. E. B. Dynkhin, *Theory of Markov processes*, Pergamon Press, New York, 1961.

22. H. Furstenberg, *Stationary processes and prediction theory*, Princeton University Press, Princeton, N. J., 1960.

23. D. Gabor, *Electronic inventions and their impact on civilization*, Imperial College of Science and Technology, London, March, 1959.

24. R. Gangolli, *Wide sense stationary sequence of distributions in Hilbert space and the factorization of operator valued functions*, J. Math. Mech. 12 (1963), 893–910.

25. P. Halmos, *Shifts of Hilbert spaces*, Crelle's Journal 208 (1961), 102–112.

26. O. Hanner, *Deterministic and non-deterministic stationary random processes*, Ark. Mat. 1 (1950), 161–177.

27. H. Helson and D. Lowdenslager, *Prediction theory and Fourier series in several variables*, Part I, Acta Math. 99 (1959) 165–202; Part II, **106** (1962), 175–213.

28. ———, *Vector-valued processes*, Proc. of the Fourth Berkeley Symposium on Mathematical Statistics and Probability, Univ. of Calif. Press, Berkeley, Calif. 1961, 203–212.

29. E. Hopf, *Ergoden Theorie*, Springer-Verlag, Berlin, 1931.

30. K. Ito, *Stationary random distributions*, Mem. Coll. of Sci. Univ. Kyoto Ser. A 28 (1953), 209–223.

31. Jang Ze-Pei, *The prediction theory of multivariate stationary processes*. I, Chinese Math. 13 (1963), 291–322.

32. S. Kakutani, Review of: *Extrapolation, interpolation, and smoothing of stationary time series*, by N. Wiener, Bull. Amer. Math. Soc. 56 (1950), 376–378.

33. ———, *Determination of the spectrum of the flow of a Brownian motion*, Proc. Nat. Acad. Sci. U.S.A. **36** (1950), 319–333.

34. R. E. Kalman and R. S. Bucy, *New results in linear filtering and prediction theory*, J. of Basic Engineering, March, 1961, 95–108.

35. K. Karhunen, *Über lineare Methoden in der Warhrscheinlickeitsrechnung*, Ann. Acad. Sci. Fenn. Ser. A I **37** (1947).

36. A. Y. Khinchin, *Korrelationstheorie der stätionaren stochastischen Prozesse*, Math. Ann. 109 (1934), 604–615.

37. A. N. Kolmogorov, *Foundations of the theory of probability*, Chelsea, New York, 1950.

38. ———, *Curves in Hilbert space which are invariant with respect to a one-parameter group of motions*, Dokl. Akad. Nauk SSSR **26** (1940), 6–9.

39. ———, *Interpolation and extrapolation of stationary random sequences*, Izv. Akad. Nauk SSSR Ser. Mat. 5 (1941), 3–14.

40. ———, *Stationary sequences in Hilbert space*, Bull. Math. Univ. Moscow **2** (1941), no. 6.

41. P. Lax, *Translation invariant spaces*, Acta Math. **101** (1959), 163–178.

42. ———, *Translation invariant spaces*, Proc. of the International Symposium on Linear Spaces, Israel Acad. Sci. Jerusalem (1961), 299–306.

43. P. Levi, *Théorie de l'addition des variables aléatoires*, Gauthier-Villars, Paris, 1937.

44. M. Loève, *Fonctions aléatoires du second ordre*, Note in P. Levi's Processus stochastiques et mouvement Brownien, Gauthier-Villars, Paris, 1948.

45. L. Mandel, *Fluctuations of light beams*, Progress of optics, Vol. II, New York, 1963, 183–248.

46. P. Masani, *The Laurent factorization of operator-valued functions*, Proc. London Math. Soc. **6** (1956), 59–69.

47. ———, *Cramer's Theorem on monotone matrix-valued functions and the Wold Decomposition*, Probability and Statistics, The Harald Cramer Volume, Wiley, New York, 1959, 175–189.

48. ———, *The prediction theory of multivariate stochastic processes*. III, Acta Math. **104** (1960), 141–162.

49. ———, *Shift invariant spaces and prediction theory*, Acta Math. **107** (1962), 275–290.

50. ———, *Sur la fonction génératrice d'un processus stochastique vectoriel*, C. R. Acad. Sci. Paris 249 (1959), 360–362.

51. ———, *Isomorphie entre les domaines temporel et spectral d'un processus vectoriel regulier*, C. R. Acad. Sci. Paris 249 (1959), 496–498.

52. ———, *Sur les fonctions matricielles de la classe de Hardy H_2*, C. R. Acad. Sci. Paris 249 (1959), 873–875; 906–907.

53. ——, *Une generalisation pour les fonctions matricielles de la classe de Hardy H_2 d'un theoreme de Nevanlinna*, C. R. Acad. Sci. Paris **251** (1960), 318–320.

54. ——, *On isometric flows on Hilbert space*, Bull. Amer. Math. Soc. **68** (1962), 624–632.

55. P. Masani and J. B. Robertson, *The time-domain analysis of continuous-parameter, weakly stationary stochastic processes*, Pacific J. Math. **12** (1962), 1361–1378.

56. R. F. Matveev, *On multidimensional regular stationary processes*, Theor. Probability Appl. **6** (1961), 149–165.

57. R. Nevanlinna, *Eindeutige analytische Funktionen*, (Zweite Auflage), Berlin, 1953.

58. J. B. Robertson, *Multivariate continuous parameter weakly stationary stochastic process*, Thesis, Indiana University, 1963.

59. M. Rosenberg, *The square-integrability of matrix-valued functions with respect to a non-negative hermitian measure*, Duke Math. J. **31** (1964), 291–298.

60. Iu. A. Rosanov, *The spectral theory of multi-dimensional stationary random processes with discrete time*, Uspehi Mat. Nauk, **13** (1958), 93–142. (Russian)

61. ——, *On the extrapolation of generalized stationary random processes*, Theory Probability Appl. **4** (1959), 465–472. (Russian)

62. ——, *Interpolation of stationary processes with discrete time*, Dokl. Acad. Nauk SSSR **130** (1960), 730–733 = Soviet Math. Dokl. **1** (1960), 91–94.

62'. ——, *Stationary random processes*, Moscow, 1963. (Russian)

63. M. Rosenblatt, *Stationary processes as shifts of functions of independent random variables*, J. Math. Mech. **8** (1959), 665–681.

64. B. B. Rossi, *Optics*, Addison-Wesley, Reading, Mass., 1957.

65. P. Whittle, *The analysis of multiple stationary time series*, J. Roy. Statist. Soc. Ser. B **15** (1953), 125–139.

66. H. Wold, *A study in the analysis of stationary time series*, 2nd ed. Almquist and Wilksell, Stockholm, 1954.

67. A. M. Yaglom, *On the problem of linear interpolation of stationary random sequences and processes*, Uspehi Mat. Nauk **4** (1949), 173. (Russian)

68. ——, *Extrapolation, interpolation and filtering of stationary random processes with rational spectral densities*, Trudy Moskov Mat. Obšč. **4** (1955), 333–374. (Russian)

69. ——, *Some classes of random fields in n-dimensional space, related to stationary random processes*, Theor. Probability Appl. **2** (1957), 273.

70. ——, *Effective solutions of linear approximation problems for multivariate stationary processes with a rational spectrum*, Theor. Probability Appl. **5** (1960), 239–264.

71. ——, *An introduction to the theory of stationary random functions*, Prentice-Hall, Englewood Cliffs, N. J., 1962.

72. L. A. Zadeh and J. R. Ragazzini, *An extension of Wiener's Theory of Prediction*, J. Appl. Phys. **21** (1950), 645–655.

73. V. Zasuhin, *On the theory of multidimensional stationary random processes*, Dokl. Akad. Nauk SSSR **33** (1941), 435–437. (Russian)

INDIANA UNIVERSITY

CONTRIBUTIONS OF NORBERT WIENER TO COMMUNICATION THEORY

BY WILLIAM L. ROOT

Information theory, communication theory, detection theory are terms now commonly used by engineers who are concerned with radio or telephone communications, electromagnetic and sonic measurements, seismic measurements, and indeed the transformation, storage and transmittal of data from any source by electronic or mechanical devices. They are terms also used by scientists concerned with the function of the human brain and nervous system, the behavior of small groups of people, the interplay between men and machines, and the behavior of whole societies. Coupled with the study of transmittal of information is the study of use of information in mechanisms, and at least one direction this latter study leads is toward the modern theory of control. Norbert Wiener was interested, as is well known, in this whole broad expanse of problems; he made this interest completely explicit in his invention of the word cybernetics, and his explication of its meaning in the book *Cybernetics* [138].* One of Professor Wiener's contributions to the scientific society was his recognition and insistence that communication and control problems which occur in very different contexts often have the same essentials; or, stated differently, that there was potentially a science of communication and control. However, it is beyond the scope of these remarks, and beyond my competence, to talk about his work from such a general point of view; so the comments here shall refer only to communication theory as it applies to electrical engineering technology.

In *I am a mathematician* [177] Wiener says "One interesting problem which we attacked together was that of the conditions restricting the Fourier transform of a function vanishing on the half line. This is a sound mathematical problem on its own merits, and Paley attacked it with vigor, but what helped me and did not help Paley was that it is essentially a problem of electrical engineering." The solution to this problem for functions of class L_2 is contained in Theorem XII of *Fourier transforms in the complex domain* [92], where it ostensibly plays the role of a key result in the theory of quasi-analytic functions. This theorem was explicitly pointed out to engineers, and its implica-

---

* The bold-faced numbers in brackets refer to numbered references in the Bibliography of Norbert Wiener. Bold-faced numbers in parentheses refer to the References at the end of this article.

tions to electric circuit synthesis discussed, by H. Wallman in the M.I.T. Radiation Laboratory Series fourteen years later (1). It is a theorem very well known to mathematical analysts, but it also became, many years after its statement, very well known to electrical engineers. It might almost be called the fundamental theorem of linear electric networks instead of the fundamental theorem of quasi-analytic functions, as Wiener dubbed it. Its significance to engineering is the following: a linear, time-invariant system in a known state at $t=0$ can be characterized by its response to an impulse at $t=0$, or by the Fourier transform of that impulse response. The Fourier transform of the impulse response is known as the transfer function of the system and is a valuable concept for the engineer, both because it is useful in calculations, and because it is useful in designing and interpreting experiments. Its complex value at a particular frequency represents the gain and phase shift of the system at that frequency. The Paley-Wiener theorem gives a necessary and sufficient condition that a given gain function be possible for a linear, time-invariant system that is *realizable*, that is, that operates only on the past. It is remarkable that Wiener knew at the time he and Paley worked on this theorem that it was electrical engineering, because few electrical engineers at that time would have recognized it as electrical engineering.

This one theorem and its history of application strikes me as typical of a large part of Wiener's contribution to the development of modern communication theory. Whether in all such instances he knew he was providing mathematical structure for communications theorists and engineers ten, twenty or thirty years hence, I do not know, but he evidently often hoped that he was. He did, of course, receive early orientation in problems of electrical engineering in discussions with Vannevar Bush in the 1920's, and these discussions must have been unusually sophisticated by the engineering standards of that time. His collaboration with Y. W. Lee somewhat later, and contacts with various electrical engineers from the time of World War II onward kept strong his ties with electrical engineering.

In a communication theory problem, man-made signals must always be of finite duration, and often their duration is limited by a fixed bound. An electrical signal of finite duration has finite energy. Hence, if one is concerned with an harmonic analysis of such signals in the communication system, the Paley-Wiener theorem (Theorem X, op. cit.) characterizing the class of functions

$$F(z) = \int_{-A}^{A} f(u)e^{iuz}\, du,$$

where $f(u)$ belongs to L_2 over $(-A, A)$ is basic. It also applies, of course, to the structural characterization of linear systems with *finite memory*, i.e., systems whose response to any input goes to zero within some interval τ after the stimulus is removed. The theorem has perhaps been less useful in application than the one cited first; I think this is so because the condition is a function-theoretic one instead of an integrability one, and therefore farther removed from the kind of constraints that appear "naturally," but this difficulty would seem to be inherent and unavoidable.

In the introduction to *Generalized harmonic analysis* [73] Wiener says, "The two theories of harmonic analysis embodied in the classical Fourier series development and the theory of Plancherel do not exhaust the possibilities of harmonic analysis. The Fourier series is restricted to the very special class of periodic functions, while the Plancherel theory is restricted to functions which are quadratically summable, and hence tend on the average to zero as their argument tends to infinity. Neither is adequate for the treatment of a ray of white light which is supposed to endure for an indefinite time."

The reference here is obviously to harmonic analysis in an L_2 context; with that limitation one may say that the Fourier transform yields an harmonic description of functions of finite energy, and the Fourier series yields an harmonic description of a very special class of functions of finite power (i.e., infinite total energy, but finite energy per unit interval). What was needed and what Wiener developed was an harmonic description of a wide class of functions of finite power, including the classes of periodic and almost periodic functions, but also including functions with a continuous or mixed "power spectrum."

The class of functions to which Wiener's generalized harmonic analysis applies is that of those measurable functions $f(t)$ for which

$$(1) \qquad \phi(\tau) = \lim_{T \to \infty} \frac{1}{2T} \int_{-T}^{T} f(\tau + t)\overline{f(t)}\, dt$$

exists for every τ. This restriction has interesting heuristic implications; for $\phi(0)$ to exist means obviously that the indefinitely continuing function $f(t)$ has associated with it an average power; for $\phi(\tau)$ to exist in general would indicate that, almost all of the time, $f(t)$ exhibits some kind of average regularity in its behavior. One would expect that many physical phenomena that are random and unpredictable in their course, but appear in an environment for which the grosser attributes are unchanging, would be appropriately modeled by such functions. Experience has borne this out, of course. One very important example is electrical noise as generated by a resistor at constant temperature, or by a vacuum tube or transistor under constant operating conditions. Since electrical noise is one of the chief ingredients in any continuous-time electrical communication problem, Wiener's generalized harmonic analysis is a fundamental tool in the analysis of such problems. It has to a certain extent been replaced formally in the analysis of electrical noise by the introduction of the stationary stochastic process as a mathematical model for noise phenomena. The theorem of Bochner that a positive-definite function is the Fourier-Stieltjes transform of a bounded monotone function, and the theorem of Khintchin that autocorrelation functions of stationary processes are positive-definite functions then provides an harmonic analysis which fits pretty much the same problems as the Wiener generalized harmonic analysis. Since the Birkhoff ergodic theorem guarantees that for almost every sample path of a metrically transitive stationary stochastic process the limit $\phi(\tau)$ as given in (1) exists and equals the autocorrelation function of the process evaluated at τ, the Wiener theory and the stationary process theory appear to one interested in physical applications to be equivalent in a strong sense. To apply the Wiener theory one has to represent the physical phenomena by a function $f(t)$ such that $\phi(\tau)$ exists; to apply the stochastic process theory one has to represent the physical phenomena by a stationary, metrically-transitive process. The extra-mathematical arguments to justify these two models seem pretty much the same. The Wiener theory came first; and it retains interest, I feel, even in applications, precisely because it deals with a single function instead of an ensemble of functions.

Electrical noise is usually such that it not only admits being represented as a stationary stochastic process, but more specifically as a stationary Gaussian stochastic process. This is true because the noise (voltage or current) is generally a macroscopic manifestation of the result of a great deal of independent activity at a microscopic level. The basic theory of Brownian motion developed by Wiener is indispensable for a meaningful study of stationary Gaussian processes with finite variance; the spectral representation of Cramer is in terms of Brownian motion for the Gaussian case; the really workable models for practical calculations involving Gaussian noise are "filtered white noise" processes, i.e., processes of moving averages,

$$(2) \qquad x(t) = \int_{-\infty}^{\infty} k(t - u)d\xi(u),$$

where $\xi(u)$ is the Brownian motion process. Modern "noise theory" and "detection theory" are applied mathematical disciplines resting chiefly (for their mathematical part) on parts of the theories of stochastic processes, harmonic analysis and statistical inference. For the first two of these Wiener's contributions are a primary source.

I have made this brief reference to some of Wiener's pure mathematics, because its influence on contemporary electrical communication engineering is perhaps greater than most mathematicians realize. I would like to distinguish between this *applicable* mathematics and a certain body of his work, mostly of later date, which seems to be more properly *applied mathematics* (or even theoretical engineering). It is a somewhat arbitrary distinction and has to do not only with subject matter but with style of writing, the presumed class of readers, and the emphasis or lack of emphasis on rigor.

The book *Extrapolation, interpolation, and smoothing of stationary time series* [144] published in 1949, but originally written as a wartime report to the National Defense Research Committee, is a borderline case, but it seems to me to belong to the second category of applied mathematics. It was written primarily for electrical engineers, even though most electrical engineers of the 1940's could not read it. This was perhaps not a bad thing; the results were sufficiently interesting to provide a stimulus to engineers to acquaint themselves with enough mathematics so that they could read it. In 1965 most theoretical communications and control engineers can read it, and in fact are fairly conversant with its contents. There is now probably not a graduate program in control and communications engineering in the country in which the material of this book is not offered. Wiener lost priority to A. N. Kolmogorov for the essential mathematical results of the prediction theory (see the footnote p. 59, op. cit., for Wiener's own comment on this), but his work was done independently of Kolmogorov's and it was addressed more explicitly to the filtering problems of electrical engineering.

One kind of problem to which the Kolmogorov-Wiener theory applies may be stated in general terms as follows: let $s(t)$, $-\infty < t < \infty$, be an intelligence-bearing signal, $n(t)$, $-\infty < t < \infty$, a noise which interferes with $s(t)$ so that an observer has available only $y(t) = s(t) + n(t)$. Suppose the observer wants a time-realizable, running estimate of $s(t)$; that is, for each t he wants $\hat{s}(t) = L(t)y_t$, where the function $\hat{s}$ should be close to s by some criterion, where

$$y_t(u) = y(u), \quad u \leq t$$
$$= 0, \qquad u > t,$$

and where $L(t)$ is a functional defined on the class $\{y_t\}$ for a suitably large class of functions y. The problem is to find a set of functions $L(t)$, $-\infty < t < \infty$, which makes $\hat{s}$ close to s. This is the kind of problem that occurs for example in designing a receiver for radio transmissions; the receiver is an apparatus which implements the operations $L(t)$.

In the original Wiener solution to this basic "smoothing" problem it is stipulated that: (1) Both $s(t)$ and $n(t)$ be functions satisfying (1), so that each has an autocorrelation function, and further that the *cross-correlation function*

$$\phi_{sn}(\tau) = \lim_{T=\infty} \frac{1}{2T} \int_{-T}^{T} s(t+\tau)n(t)dt$$

exists for each τ. These correlation functions are to be known in advance.

(2) The operations $L(t)$ are to be linear operations,

$$\hat{s}(t) = L(t)y_t = \int_{0}^{\infty} y(t-\tau)dK(\tau)$$

which are time-invariant.

(3) The criterion for the closeness of $\hat{s}(t)$ and $s(t)$ is the mean-squared-error,

$$(3) \qquad \lim_{T\to\infty} \frac{1}{2T} \int_{-T}^{T} \left| s(t) - \int_{0}^{\infty} y(t-\tau)dK(\tau) \right|^2 dt.$$

The solution for $K(\tau)$, or rather for its Fourier-Stieltjes transform, is given in Equation (3.28) *op. cit.* The solution is formally the same if condition (1) is replaced by the condition that $s(t)$, $n(t)$ and $(s(t)$, $n(t))$ are stationary stochastic processes and the time-average auto and cross-correlation functions are replaced by the corresponding statistical averages, as, for example,

$$\lim_{T\to\infty} \frac{1}{2T} \int_{-T}^{T} s(t+\tau)\overline{s(t)}dt \sim Es(t+\tau)s(t),$$

where E denotes mathematical expectation, and if the criterion (3) is replaced by

$$E \left| s(t) - \int_{0}^{\infty} y(t-\tau)dK(\tau) \right|^2$$

for each t. Further, if $s(t)$, $n(t)$ and $(s(t)$, $n(t))$ are metrically transitive, the solution is exactly the same for each pair of sample functions $s(t)$, $n(t)$ except for those in a set of probability zero.

Modern communication and signal detection theory began when it was realized that a meaningful way to treat such a problem was to treat it statistically, to ascribe to $s(t)$ and $n(t)$ certain average properties, rather than to specify them explicitly, and to use some kind of statistical or average criterion to judge the closeness of $\hat{s}$ to s. There is, of course, a great variety of changes on such a problem; the perturbation may not be caused by simple additive noise, the structure of possible signals $s(t)$ may be more or less narrowly specified, the *a priori* statistical information available may vary widely in kind and quantity, the statistical criteria need not be mean-squared-error, etc. The statistical point of view had gained some acceptance during and shortly after the war, due largely to the work of a dozen or so people, some of whom were mathematicians and physicists who had been led to think of radio communication (and measurement) problems because of their wartime jobs. I will venture the opinion that there were three papers or monographs published in this country in the middle and late 1940's that were more than any others responsible for establishing statistical ratio communications (and radio measurement) theory; these are the papers on random noise by S. O. Rice (2), on information theory by C. Shannon (3), and the monograph on smoothing and predicting by Wiener [144].

The original theory of filtering and prediction was linear (although Wiener also wrote on nonlinear prediction, see [170], [196]). It fitted neatly onto Wiener's earlier work on generalized harmonic analysis and his work with E. Hopf on a class of linear integral equations [78]. But Wiener was also interested in nonlinear problems in enginering; he wanted to provide a rather general structure of theory to support problems in the synthesis and analysis of nonlinear electrical networks and other nonlinear systems. Apparently most of what he did in this area was done in the 1940's, but he published only one report, *Response of a Nonlinear Device to Noise*, M.I.T. Radiation Laboratory Report No. 129, 1942, and no papers in the scientific journals. The first really open publication was the set of lectures, *Nonlinear Problems in Random Theory* [191], in 1958. Some of this work was done in collaboration with Y. W. Lee, and initiated a considerable body of engineering research by Lee and his students in the years after 1950.

Wiener considered nonlinear systems whose output at a particular instant when driven by Brownian motion could be written as functionals of the Brownian motion of the form

$$(4) \quad y(\alpha) = K_0 + \sum_{n=1} \int \cdots \int K_n(\tau_1, \cdots, \tau_n) \, dx(\tau_1, \alpha) \cdots dx(\tau_n, \alpha)$$

where the kernels K_n are characteristic of the system and where $x(\tau, \alpha)$ is Brownian motion. If the output at a time t seconds later is represented by the same functional except with the arguments of the K_n translated by t, this expression gives the running output of a time-invariant nonlinear system. If two systems yield the same output $y(\alpha)$ when excited by $x(\tau, \alpha)$ for all α in a set of probability one, they are equivalent in a strong sense; they would certainly seem to be equivalent for all engineering purposes. An engineering analysis problem, then, is to find the K_n's for a given system specified in some arbitrary way; a synthesis problem is, e.g., to specify an electrical network which will realize a given set of K_n's. Wiener attacked these problems by first orthogonalizing the sequence of homogeneous polynomial integral functionals appearing in (4) into a sequence of non-homogeneous polynomial integral functionals $G_n(K_n, \alpha)$ with the property that $G_n(K_n, \alpha)$ is orthogonal to $G_m(H_m, \alpha)$, $m \neq n$, for any kernel H_m, with respect to Wiener measure. Then any functional of the form of (4) can be written

$$(5) \qquad\qquad y(\alpha) = \sum_n G_n(K_n', \alpha)$$

for an appropriate choice of kernels K_n', where now the summands are orthogonal in a statistical sense, i.e., with respect to Wiener measure.

This work of Wiener and Lee (see also (4)) together with the contributions of three or four others largely set the direction that has been followed since for most of the investigations of a general theory of nonlinear communication and control systems with stochastic inputs. Usually the computational complexities arising when one tries to apply such general theory to a specific problem are so great that, thus far, the practical results are limited. Ad hoc procedures are often more effective. There is hope, of course, that electronic computation will eventually overcome these difficulties.

Finally I should like to note that apart from the theorems he proved and the calculations he made, Wiener performed one other service for electrical communication engineering: he was an effective propagandist for the new ideas of the 1940's about the statistical nature of information and its transmission. Because of his prestige as a mathematician, but more than that, because of his reputation amongst engineers as a first rank mathematician aware of electrical engineering problems, he was listened to when he made statements such as the following, from the magazine, *Electronics*, 1949 [145]: "There is no vagueness in the definition of energy, so that power engineering is a field in which the objectives have been fully understood for a long time. On the other hand, most books on communications say nothing about information, and the average communication engineer does not have a definite measure of information. He studies communication circuits as they are affected by sinusoidal inputs, but he does not discuss the relation between sinusoidal and information-carrying inputs. Only within the past few years have a few engineers begun asking what information is and using the concept of information as a basis for design."

"Because information depends, not merely on what is actually said, but on what might have been said, its measure is a property of a set of possible messages, or of what is called an *ensemble* in statistical mechanics. Such as ensemble is more than a set taken simply; it is a set to which is attributed a probability measure. We thus have a situation that is closely akin to that in statistical mechanics, more especially in the form which Gibbs gave it."

References

1. G. R. Valley and H. Wallman, *Realizability of filters*, Appendix A, Vacuum tube amplifiers, MIT Radiation Laboratory Series, Vol. 18, McGraw-Hill, New York, 1948.

2. S. O. Rice, *Mathematical analysis of random noise*, Bell System Tech. J. 23 (1944), 282–332; 24 (1945), 46–156.

3. C. E. Shannon, *A mathematical theory of communications*, Bell System Tech. J. 27 (1948), 379–423; 27 (1948), 623–656.

4. Y. W. Lee, *Contributions of Norbert Wiener to linear theory and nonlinear theory in engineering* in Selected Papers of Norbert Wiener, S.I.A.M. and M.I.T. Press, Cambridge, 1964.

The University of Michigan

BIBLIOGRAPHY OF NORBERT WIENER

1. *On the rearrangement of the positive integers in a series of ordinal numbers greater than that of any given fundamental sequence of omegas*, Messenger of Math. **3** (1913), No. 511.

2. *The highest good*, J. Phil. Psych. and Sci. Method **9** (1914), 512–520.

3. *Relativism*, J. Phil. Psych. and Sci. Method **9** (1914), 561–577.

4. *A simplification of the logic of relations*, Proc. Cambridge Philos. Soc. **27** (1914), 387–390.

5. *A contribution to the theory of relative position*, Proc. Cambridge Philos. Soc. **27** (1914), 441–449.

6. *Studies in synthetic logic*, Proc. Cambridge Philos. Soc. **18** (1915), 24–28.

7. *The shortest line dividing an area in a given ratio*, J. Phil. Psych. and Sci. Method (1915), 567–574.

8. *Certain formal invariance in Boolean algebras*, Trans. Amer. Math. Soc. **18** (1917), 65–72.

9. *Bilinear operations generating all operations rational in a domain*, Ann. of Math. **21** (1920), 157–165.

10. *A set of postulates for fields*, Trans. Amer. Math. Soc. **21** (1920), 237–246.

11. *Certain iterative characteristics of bilinear operations*, Bull. Amer. Math. Soc. **27** (1920), 6–10.

12. *The mean of a functional of arbitrary elements*, Ann. of Math. (2) **22** (1920), 66–72.

13. *On the theory of sets of points in terms of continuous transformations*, G. R. Strasbourg Math. Congress, 1920.

14. *Certain iterative properties of bilinear operations*, G. R. Strasbourg Math. Congress, 1920.

15. *A new theory of measurement: A study in the logic of mathematics*, Proc. London Math. Soc. **19** (1921), 181–205.

16. *A new vector in integral equations* (with F. L. Hitchcock), J. Math. and Phys. **1** (1921), 20 pp.

17. *The average of an analytical functional*, Proc. Nat. Acad. Sci. U.S.A. **7** (1921), 253–260.

18. *The average of an analytical functional and the Brownian movement*, Proc. Nat. Acad. Sci. U.S.A. **7** (1921), 294–298.

19. *The isomorphisms of complex algebra*, Bull. Amer. Math. Soc. **27** (1921), 443–445.

20. *The relation of space and geometry to experience*, Monist. **32** (1922), 12–60; 200–247; 364–394.

21. *The group of the linear continuum*, Proc. London Math. Soc. 20 (1922), 329–346.

22. *A new type of integral expansion*, J. Math. and Phys. 1 (1922), 167–176.

23. *Limit in terms of continuous transformation*, Bull. Soc. Math. France (1922), 119–134.

24. *Certain notions in potential theory*, J. Math. and Phys. 3 (1924), 24–51.

25. *The equivalence of expansions in terms of orthogonal functions* (with J. L. Walsh), J. Math. and Phys. 1 (1922), 103–122.

26. *On the nature of mathematical thinking*, Austral. J. Psych. and Phil. 1 (1923), 268–272.

27. *Note on a paper of M. Banach*, Fund. Math. 4 (1923), 136–143.

28. *Nets and the Dirichlet problem* (with H. B. Phillips), J. Math. and Phys. 2 (1923), 105–124.

29. *Differential space*, J. Math. and Phys. 2 (1923), 131–174.

30. *Note on a new type of summability*, Amer. J. Math. 45 (1923), 83–86.

31. *Discontinuous boundary conditions and the Dirichlet problem*, Trans. Amer. Math. Soc. 25 (1923), 307–314.

32. *Note on the series* $\sum(+1/n)$, Bull. Acad. Polon. Ser. A (1923), 87–90.

33. *In memory of Joseph Lipka*, J. Math. and Phys. 3 (1924), 63–65.

34. *The quadratic variation of a function and its Fourier coefficients*, J. Math. and Phys. 3 (1924), 72–94.

35. *The Dirichlet problem*, J. Math. and Phys. 3 (1924), 127–147.

36. *Une condition nécessaire et suffisante de possibilité pour le problème de Dirichlet*, C. R. Acad. Sci. Paris 178 (1924), 1050–1053.

37. *The average value of a functional*, Proc. London Math. Soc. 22 (1924), 454–467.

38. *Un problème de probabilités dénombrables*, Bull. Soc. Math. France 11 (1924), 3–4.

39. *Note on a paper of O. Perron*, J. Math. and Phys. 4 (1925), 21–32.

40. *The solution of a difference equation by trigonometrical integrals*, J. Math. and Phys. 4 (1925), 153–163.

41. *A contribution to the theory of interpolation*, Ann. of Math. (2) 26 (1925), 212–116.

42. *Note on quasi-analytic functions*, J. Math. and Phys. 4 (1925), 193–199.

43. *On the representation of functions by trigonometrical integrals*, Math. Z. 24 (1925), 576–616.

44. *Verallgemeinerte Trigonometrische Entwicklungen*, Gott. Nachrichten (1925), 151–158.

45. *A new formulation of the laws of quantization for periodic and aperiodic phenomena* (with M. Born), J. Math. and Phys. **5** (1926), 84–98.

46. *Eine neue Formulierung der Quantengesetze fur Periodische und nicht Periodische Borganze*, Z. Physik **36** (1926), 174–187.

47. *The harmonic analysis of irregular motion*, J. Math. and Phys. **5** (1926), 99–121.

48. *The harmonic analysis of irregular motion* (Second Paper), J. Math. and Phys. **5** (1926), 158–189.

49. *The operational calculus*, Math. Ann. **95** (1926), 557–584.

50. *Analytical approximations to topological transformations* (with P. Franklin), Trans. Amer. Math. Soc. **28** (1926), 762–785.

51. *On the closure of certain assemblages of trigonometrical functions*, Proc. Nat. Acad. Sci. **13** (1927), 27–29.

52. *Laplacians and continuous linear functionals*, Acta Sci. Math. (Szeged) **3** (1927), 7–16.

53. *The spectrum of an array and its application to the study of the translation properties of a simple class of arithmetical functions*, J. Math. and Phys. **6** (1927), 145–157.

54. *Quantum theory and gravitational relativity* (with D. J. Struik), Nature **119** (1927), 853–854.

55. *A relativistic theory of quanta* (with D. J. Struik), J. Math. and Phys. **7** (1927), 1–23.

56. *Sur la théorie relativiste des quanta* (with D. J. Struik), C. R. Acad. Sci. Paris **185** (1927), 42–44.

57. *On a new definition of almost periodic functions*, Ann. of Math. (2) **28** (1927), 365–367.

58. *On a theorem of Bochner and Hardy*, J. London Math. Soc. **2** (1927), 118–123.

59. *Sur la théorie relativiste des quanta*, C. R. Acad. Sci. Paris **185** (1927), 184–185.

60. *Une généralisation des fonctionelles a variation bornée*, C. R. Acad. Sci. Paris **185** (1927), 65–67.

61. *Une méthode nouvelle pour la démonstration des théorèms de Tauber*, C. R. Acad. Sci. Paris **184** (1927), 793–795.

62. *The fifth dimension in relativistic quantum theory* (with D. J. Struik), Proc. Nat. Acad. Sci. **14** (1928), 262–268.

63. *The spectrum of an arbitrary function*, Proc. London Math. Soc. **27** (1928), 287–496.

64. *Coherency matrices and quantum theory*, J. Math. and Phys. **1** (1928), 109–125.

65. *A new method of Tauberian theorems*, J. Math. and Phys. **7** (1928), 161–184.

138 BIBLIOGRAPHY

66. *Harmonic analysis and the quantum theory*, J. Franklin Inst. **207** (1929), 525–534.

67. *Hermitian polynomials and Fourier analysis*, J. Math. and Phys. **8** (1929), 70–73.

68. *Harmonic analysis and group theory*, J. Math. and Phys. **8** (1929), 148–154.

69. *On the spherically symmetrical statistical field in Einstein's unified theory of electricity and gravitation*, Proc. Nat. Acad. Sci. **15** (1929), 353–356.

70. *On the spherically symmetrical statistical field in Einstein's unified theory: A correction* (with M. S. Vallarta), Proc. Nat. Acad. Sci. **15** (1929), 802–804.

71. *A type of Tauberian theorem applying to Fourier series*, Proc. London Math. Soc. **30** (1929), 1–8.

72. *Fourier analysis and asymptotic series*, Operational circuit analysis by V. Bush, Wiley, New York, 1929; Appendix, pp. 366–379.

73. *Generalized harmonic analysis*, Acta Math. **55** (1930), 117–258.

74. *Tauberian theorems*, Ann. of Math. **33** (1932), 1–100.

75. *A note on tauberian theorems*, Ann. of Math. **33** (1932), 787.

76. *A new deduction of the Gaussian distribution*, J. Math. and Phys. **10** (1931), 284–288.

77. *Characters of infinite Abelian groups* (with R. E. A. C. Paley), Int'l. Math. Congr., Zürich, 1932.

78. *Über eine Klasse Singulärer Integralgleichungen* (with E. Hopf), S.-B. Preuss. Akad. Wiss. (1932), 696.

79. Review: Harald Bohr. *Fastperiodische Funktionen*, Math. Gaz. **17** (1933), No. 54.

80. Review: A. S. Besicovitch. *Almost periodic functions*, Math. Gaz. **16** (1932), No. 220.

81. *The Fourier integral and certain of its applications*, Cambridge Univ. Press, New York, 1933; reprint, Dover, New York, 1959; Math. Rev. **20** (1959), rev. no. 6634.

82. *Notes on the theory and application of Fourier transforms* (with R. E. A. C. Paley) I, II, Trans. Amer. Math. Soc. **35** (1933), 348–355; III, IV, V, VI, VII, ibid. **35** (1933), 761–791.

83. *A one-sided Tauberian theorem*, Math. Z. **36** (1933), 787–789.

84. *R. E. A. C. Paley—In memoriam*, Bull. Amer. Math. Soc. **39** (1933), 476.

85. *Characters of Abelian groups* (with R. E. A. C. Paley), Proc. Nat. Acad. Sci. **19** (1933), 253–257.

86. Review: E. C. Titchmarsh, *The Fourier integral and certain of its applications*, Math. Gaz. **17** (1933), 129; Science **132** (1933), 731.

87. *Notes on random functions* (with R. E. A. C. Paley and A. Z. Zygmund), Math. Z. 37 (1933), 647–668.

88. *The total variation of $g(xh) - f(x)$* (with R. C. Young), Trans. Amer. Math. Soc. 35 (1933), 327–340.

89. *Leibnitz and Haldane*, Philos. Sci. 1 (1934), No. 4.

90. *Notes on the Kron theory of tensors in electrical machinery*, Abstract, J. Electr. Eng. China 7, 3–4.

91. *A class of gap theorems*, Ann. Scuola Norm. Sup. Pisa, E (1934–1936), 1–6.

92. *Fourier transforms in the complex domain* (with R. E. A. C. Paley), Amer. Math. Soc. Colloq. Publ., Vol. 19, Amer. Math. Soc., Providence, R. I., 1934.

93. *Random functions*, J. Math. and Phys. 14 (1934), No. 1.

94. *The closure of Bessel functions*. Abstract 66, Bull. Amer. Math. Soc. 41 (1935), 35.

95. *Fabry's gap theorem*, Sci. Repts. of Nat'l. Tsing Hua Univ., Ser. A, 3 (1935), 239–245.

96. *A theorem of Carleman*, Sci. Repts. of Nat'l. Tsing Hua Univ., Ser. A, 3 (1935), 291–298.

97. *Math. in American secondary schools*, J. Math. Assoc. Japan for Secondary Education, Tokyo, 1935.

98. *The role of the observer*, Philos. Sci. 3 (1936), 307–319.

99. *Sur les séries de Fourier lacunaires. Théorème direct* (with Szolem Mandelbrojt), C. R. Acad. Sci. Paris 203 (1936), 34–36.

100. *Séries de Fourier lacunaires. Théorème inverses*, with Szolem Mandelbrojt, C. R. Acad. Sci. Paris 203 (1936), 233–234.

101. *Gap theorems*, C. R. de Congr. Int'l. des Math., 1936.

102. *A Tauberian gap theorem of Hardy and Littlewood*, J. Chinese Math. Soc. 1 (1936), 15.

103. *Taylor's series of entire functions of smooth growth* (with W. T. Martin), Duke Math. J. 3 (1937), 213–223.

104. *Random Waring's theorems*. Abstract (with N. Levinson), Science 85 (1937), 439.

105. *On absolutely convergent Fourier-Stieltjes transforms* (with II. R. Pitt), Duke Math. J. 4 (1938), No. 2.

106. *Fourier-Stieltjes transforms and singular infinite convolutions* (with Aurel Wintner), Amer. J. Math. 60 (1938), No. 3.

107. *Remarks on the classical inversion formula for the Laplace integral* (with D. V. Widder), Bull. Amer. Math. Soc. 44 (1938), 573.

108. *The homogeneous chaos*, Amer. J. Math. 60 (1938), 897–936.

109. *The decline of cookbook engineering*, Tech. Rev. 41 (1938), 23.

110. Review: L. Hogben, *Science for the citizen*, Tech. Rev. 41 (1938), 66–67.

111. *The historical background of harmonic analysis*, Amer. Math. Soc. Semicentennial Publications Vol. II, Semicentennial Addresses, Amer. Math. Soc., Providence, R. I., 1938.

112. Review: Roger Burlingame, *March of the iron men*, Tech. Rev. **41** (1939), 115.

113. *Convergence properties of analytic functions of Fourier-Stieltjes transforms* (with R. H. Cameron), Trans. Amer. Math. Soc. **46** (1939), 97–109; Math. Rev. **1** (1940), 13; 400.

114. *Generalization of Ikehara's theorem* (with J. R. Pitt), J. Math. and Phys. **17** (1939), No. 4.

115. *On singular distribution* (with Aurel Wintner), J. Math. and Phys. **17** (1939), No. 4.

116. *A new method in statistical mechanics* (with B. McMillan), Abstract 133, Bull. Amer. Math. Soc. **45** (1939), 234; Science **90** (1939), November 3.

117. *The ergodic theorem*, Duke Math. J. **5** (1939), 1–18.

118. *The use of statistical theory in the study of turbulence*, Nature **144** (1939), 728.

119. *A canonical series for symmetric functions in statistical mechanics.* Abstract 133, Bull. Amer. Math. Soc. **46** (1940), 57.

120. Review: M. Fukamiya, *On dominated ergodic theorems in L_p ($p \geq 1$)*, Math. Rev. **1** (1940), 148.

121. Review: M. Fukamiya, *The Lipschitz condition of random functions*, Math. Rev. **1** (1940), 149.

122. Review: Th. De Donder, *L'énergétique déduite de la mécanique statistique générale*, Math. Rev. **1** (1940), 192.

123. *Harmonic analysis and ergodic theory* (with Aurel Wintner), Amer. J. Math. **63** (1941), 415–426; Math. Rev. **2** (1941), 319.

124. *On the ergodic dynamics of almost periodic systems* (with Aurel Wintner), Amer. J. Math. **63** (1941), 794–824.

125. *On the oscillation of the derivatives of a periodic function* (with George Pólya), Trans. Amer. Math. Soc. **52** (1942), 249–256.

126. *Ergodic dynamics of almost periodic systems* (with Aurel Wintner), Math. Rev. **4** (1943), 15 (same as reference 124 above).

127. *Behavior, purpose, and teleology* (with Arturo Rosenblueth and J. Bigelow), Philos. Sci. **10** (1943), 18–24.

128. *The discrete chaos* (with Aurel Wintner), Amer. J. Math. **65**, 279–298; Math. Rev. **4** (1943), 220.

129. *The role of models in science* (with Arturo Rosenblueth), Philos. Sci. **12** (1945), 316–322.

130. *La teoria de la extrapolacion estadistica*, Bol. Soc. Mat. Mexicana **2** (1945), 37–42; Math. Rev. **7** (1946), 461.

131. *The mathematical formulation of the problem of conduction of im-*

pulses in a network of connected excitable elements, specifically in cardiac muscle (with Arturo Rosenblueth), Arch. Inst. Cardiol. Méxicana **16** (1946), 205–265; Bol. Soc. Mat. Mexicana **2** (1945), 37–42; Math. Rev. **9** (1948), 604.

132. *Theory of statistical extrapolation*, Math. Rev. **7** (1946), 416 (same as reference 130 above).

133. *A generalization of the Wiener-Hopf integral equation* (with Albert E. Heins), Proc. Nat. Acad. Sci. U.S.A. **32** (1946), 98–101; Math. Rev. **8** (1947), 29.

134. *Sur les fonctions indéfiniment dérivables sur une demi-droite* (with Szolem Mandelbrojt), C. R. Acad. Sci. Paris **225** (1947), 978–980; Math. Rev. **9** (1948), 230.

135. *A scientist rebels*, Atlantic Monthly **179** (1947), 46.

136. *Teleological mechanisms* (with L. K. Frank, G. E. Hutchinson, W. K. Livingston, and W. S. McCulloch), Ann. New York Acad. Sci. **50** (1948), 187–278.

137. Review: L. Infeld, *Whom the Gods love: The story of Evariste Galois*, Scripta Math. **14** (1948), 273–274.

138. *Cybernetics, or control and communication in the animal and the machine*, Actualités Sci. Ind., no. 1053; Hermann et Cie., Paris; The MIT Press, Cambridge, Mass. and Wiley, New York, 1948; Math. Rev. **9** (1948), 598.

139. *A rebellious scientist after two years*, Bull. Atomic Scientists **4** (1948), 338.

140. *Time, communication and the nervous system*, Ann. New York Acad. Sci. **50** (1948), 197–220; Math. Rev. **10** (1949), 133.

141. *Cybernetics*, Scientific American **179** (1948), 14–18.

142. *An account of the spike potential of axons* (with Arturo Rosenblueth, W. Pitts, and J. Garcia Ramos), J. Comp. Physiol. (1948), December.

143. *Sur la théorie de la prévision statistique et du filtrage des ondes*, Analyse Harmonique, Colloques Internationaux du CNRS, No. 15, pp. 67–74. Centre National de la Recherche Scientifique, Paris, 1949; Math. Rev. **11** (1950), 376.

144. *Extrapolation, interpolation, and smoothing of stationary time series. With engineering applications*, The MIT Press, Cambridge, Mass.; Wiley, New York; Chapman & Hall, London, 1949; Math. Rev. **11** (1950), 118.

145. *New concept of communication engineering*, Electronics **22** (1949), 74–77.

146. Obituary: *Godfrey Harold Hardy* (1877–1947), Bull. Amer. Math. Soc. **55** (1949), 72–77; Math. Rev. **10** (1949), 420.

147. *Sound communication with the deaf*, Philos. Sci. **16** (1949), No. 3.

148. *Some problems in sensory prosthesis* (with L. Levine), Science **110** (1949), November 11.

149. *The thinking machine*, Time **55** (1950), January 23.

150. *Some prime-number consequences of the Ikehara theorem* (with Leonard Geller), Acta Sci. Math. (Szeged) **12**, Leopoldo Fejer et Frederico Riesz LXX annos natis dedicatus, Pars B, 25–28 (1950); Math. Rev. **11** (1950), 644; ibid **12** (1951), 1002.

151. *The human use of human beings*, Houghton Mifflin, Boston, 1950 (paperback edition (Anchor) by Doubleday, 1954).

152. *Cybernetics*, Bull. Amer. Acad. Arts and Sci. **3** (1950), No. 7.

153. *Some maxims for biologists and psychologists*, Dialectica **4** (September 1950), 3.

154. *Purposeful and non-purposeful behavior* (with Arturo Rosenblueth), Philos. Sci. (1950), October.

155. *Speech, language, and learning*, J. Acoust. Soc. Amer. **22** (1950), 696–697.

156. *Comprehensive view of prediction theory*, Proceedings of the International Congress of Mathematicians, Cambridge, Mass., 1950, Vol. 2, pp. 308–321; Amer. Math. Soc., Providence, R. I., 1952, Expository lecture; Math. Rev. **13** (1952), 477.

157. *Entropy and information*, Proc. Sympos. Appl. Math., Vol. **2**, Amer. Math. Soc., Providence, R. I., 1950; p. 89; Math. Rev. **11** (1950), 305.

158. *Problems of sensory prosthesis*, Bull. Amer. Math. Soc. **57** (1951), 27–35.

159. *Homeostasis in the individual and society*, J. Franklin Inst. **251** (1951), No. 1.

160. *The brain (ss)*, Tech. Eng. News (1952), April; reprinted in paperback anthology, Crossroads in Time, ed. Groff Conklin, Doubleday, New York, 1953.

161. *Les machines à calculer et la forme (Gestalt)*, Les machines à calculer et la pensée humaine, Colloques Internationaux du Centre National de la Recherche Scientifique, Paris, 1953; pp. 461–463; Math. Rev. **16** (1955), 529.

162. *A new form for the statistical postulate of quantum mechanics* (with Armand Siegel), Phys. Rev. **91** (1953), 1551–1560; Math. Rev. **15** (1954), 273.

163. *Distributions quantiques dans l'espace différentiel pour les fonctions d'ondes dépendant du spin*, C. R. Acad. Sci. Paris **237** (1953), 1640–1642; Math. Rev. **15** (1954), 490.

164. *Miracle of the broom closet (ss)*, Tech. Eng. News **33** (1952), No.

7; reprinted in the Magazine of Fantasy and Science Fiction, ed. Anthony Boucher, February, 1954.

165. *Ex-prodigy: my childhood and youth*, Simon and Schuster, New York, 1953; The MIT Press, Cambridge, Mass. 1965 (also paperback edition by MIT Press); Math. Rev. **15** (1954), 277.

166. *The concept of homeostasis in medicine*, Transactions and Studies of the College of Physicians of Philadelphia (4) **20** (1953), No. 3.

167. *The differential space theory of quantum mechanics* (with Armand Siegel), Phys. Rev. **91** (1953), 1551.

168. *Optics and the theory of stochastic processes*, J. Opt. Soc. Amer. **43** (1953); 225–228; Math. Rev. **17** (1956), 33.

169. *The future of automatic machinery*, Mechanical Engineering (1953), 130–132, February.

170. *Nonlinear prediction and dynamics*, Proc. Third Berkeley Symposium on Mathematical Statistics and Probability, University of California Press, 1954; Math. Rev. **18** (1957), 949.

171. *The differential space theory of quantum systems* (with Armand Siegel), Nuovo Cimento (10) **2** (1955), 982–1003, No. 4, Suppl.

172. *On the factorization of matrices*, Comment. Math. Helv. **29** (1955), 97–111; Math. Rev. **16** (1955), 921.

173. *Thermodynamics of the message*, Neurochemistry, ed. K. E. C. Elliott, Thomas, Springfield, 1955.

174. *Time and organization*, Second Fawley Foundation Lecture, University of Southampton, 1955.

175. *The "Theory of measurement" in differential space quantum theory* (with Armand Siegel), Phys. Rev. **101** (1956), 429–432.

176. *The theory of prediction*, Modern mathematics for the engineer, ed. E. F. Beckenbach, McGraw-Hill, New York, 1956.

177. *I am a mathematician. The later life of a prodigy*, Doubleday, Garden City, New York, 1956 (paperback edition by The MIT Press); Math. Rev. **17** (1956), 1037.

178. *Brain waves and the interferometer*, J. Phys. Soc. Japan **18** (1956), No. 8.

179. *Moral reflections of a mathematician*, Bull. Atomic Scientists **12** (1956), 53–57; reprinted from *I am a mathematician*.

180. *Rhythms in physiology with particular reference to encephalography*. Proceedings of the Rudolf Virchow Medical Society in the City of New York, Vol. 16, 1957; pp. 109–124.

181. *The definition and ergodic properties of the stochastic adjoint of a unitary transformation* (with E. J. Akutowicz), Rend. Circ. Mat. Palermo (2) **6** (1957), 205–217, Addendum, 349; Math. Rev. **20** (1959), rev. no. 4328.

182. *Notes on Pólya's and Turán's hypotheses concerning Liouville's*

factor, Rend. Circ. Mat. Palermo (2) **6** (1957), 240–248; Math. Rev. **20** (1959), rev. no. 5759.

183. *The role of the mathematician in a materialistic culture (A scientist's dilemma in a materialistic world)*, Columbia Engineering Quarterly, Proceedings of the Second Combined Plan Conference, Arden House, October 6–9, 1957; pp. 22–24.

184. *On the non-vanishing of Euler products* (with Aurel Wintner), Amer. J. Math. **79** (1957), 801–808.

185. *The prediction theory of multivariate stochastic processes. I. The regularity condition* (with P. Masani), Acta Math. **98** (1957), 111–150; Math. Rev. **20** (1959), rev. no. 4323.

186. *The prediction theory of multivariate stochastic processes. II. The linear predictor* (with P. Masani), Acta Math. **99** (1958), 93–137; Math. Rev. **20** (1959), rev. no. 4325.

187. *Logique, probabilité et méthode des sciences physiques*, La Methodes dans les Sciences Modernes, Editions Science et Industrie, ed. François Le Lionnais, Paris, 1958; pp. 111–112.

188. *My connection with cybernetics. Its origin and its future*, Cybernetica (1958), 1–14.

189. *Random time*, Nature **181** (1958), 561–562.

190. *Time and the science of organization*, Scientia (1958), September.

191. *Nonlinear problems in random theory*, The MIT Press, Cambridge, Mass., and Wiley, New York 1958; Math. Rev. **20** (1959), rev. no. 7337.

192. *Sur la prévision linéaire des processus stochastiques vectoriels à densité spectrale bornée. I* (with P. Masani), C. R. Acad. Sci. Paris **246** (1958), 1492–1495; Math. Rev. **20** (1959), rev. no. 4324a.

193. *Sur la prévision linéaire des processus stochastiques vectoriels à densité spectrale bornée. II* (with P. Masani), C. R. Acad. Sci. Paris **246** (1958), 1655–1656; Math. Rev. **20** (1959), rev. no. 4324b.

194. *A factorization of positive Hermitian matrices* (with E. J. Akutowicz), J. Math. Mech. **8** (1959), 111–120.

195. *The tempter*, Random House, New York, 1959.

196. *Non-linear prediction* (in "Probability and statistics") (with P. Masani), The Harald Cramer Volume, ed. U. Grenander, Stockholm, 1959.

197. *On bivariate stationary processes and the factorization of matrix-valued functions* (with P. Masani), Theory of Probability and its Applications (Moscow) **4** (1959), 322–331; (English transl., pp. 300–308).

198. Preface to "Cybernetics of natural systems" by D. Stanley-Jones, Pergamon Press, London, 1960.

199. *"The grand privilege,"* Saturday Review (1960), March 5.

200. *Some moral and technical consequences of automation*, Science **131** (1960).

201. *Kybernetik*, Contribution to *Das Soziologische Wörterbuch*, F. Enke Verlag, Stuttgart, 1960.

202. *Ueber Informationstheorie*, Naturwissenschaften **7** (1961), 174–176.

203. *Science and society*, Voprosy Filosofii (1961), No. 7.

204. *Cybernetics*, Second edition, The MIT Press and Wiley, New York, 1961, (also paperback edition by MIT Press).

205. *L'homme et la machine*, Proc. Colloques Philosophiques Internationaux de Royaumont, July, 1962; *Le concept d'information dans la science contemporaine*, Gauthier-Villars, Paris, 1965, pp. 99–132.

206. Contribution to Proc. of the International Symposium on the Application of Automatic Control in Prosthetics Design, August 27–31, 1962, Opatija, Yugoslavia; p. 132.

207. *The mathematics of self-organizing systems*, Recent developments in information and decision processes, Macmillan, New York, 1962.

208. *Introduction to neurocybernetics and Epilogue* (With J. P. Schade), Progress in brain research, Vol. 2 (Nerve, Brain and Memory Models), Elsevier Publishing Co., Amsterdam, 1963, pp. 1–7, 264–268.

209. *Random theory in classical phase space and quantum mechanics* (with Giacomo della Riccia), Proc. Internat. Conference on Functional Analysis, Massachusetts Institute of Technology, Cambridge, Mass., June 9–13, 1963; Analysis in function space, pp. 3–14, The MIT Press, Cambridge, Mass. 1964.

210. *On the oscillations of nonlinear systems*, Proc. Symposium on Stochastic Models in Medicine and Biology, Mathematics Research Center, U. S. Army, June 12–14, 1963.

211. *God and Golem, Inc.: A comment on certain points where cybernetics impinges on religion*, The MIT Press, Cambridge, Mass., 1964.

212. *Dynamical systems in physics and biology*, Contribution to series "Fundamental Science in 1984," The New Scientist, London, 1964.

213. Reprinting of [143] as *Time series*, Wiley paperbacks, 1964 (paperback edition by The MIT Press).

214. *Selected papers of Norbert Wiener*: Expository papers by Y. W. Lee, Norman Levinson, and W. T. Martin, The MIT Press, Cambridge, Mass., 1964.